CROSSING INFINITY

THE COLLAPSE OF CIVILIZATION AS WE KNOW IT

ETC

CROSSING INFINITY

DISTILLING

The Great Disruption by Paul Gilding

The Upside of Down by Thomas Homer-Dixon

Coming Apart by Charles Murray

Collapse by Jared Diamond

Black Swan by Nassim Taleb

The World is Flat by Thomas Fiedman

TOM WEATHERS

Gastonia nc

possumgolightly press

ISBN 978-1-304-02627-9

Published by Possumgolightly Press

Gastonia, NC USA

possumgolightly.com

CONTENTS

DISTILLATIONS

Possumgolightly distillations organize the basic facts of important
books - finding patterns that the original author might not have seen.

CROSSING INFINITY

Lines and curves approach but never touch.
The lines push the curves from side to side from bottom to top
toward the giddy heights
and ultimate consumption
almost straight up
then no longer balanced
falling into the bottomless pit
crossing infinity

THEME(S) AND ASYMPTOTES

STARTING WITH TWO ASIDES
We need to get this out of the way.

Why?

These are important connected books - collectively maybe the most important books around. I don't think anybody has ever seen the connection. Probably not even the writers. Unlikely as it seems, I might be the first. The themes below reveal these heretofore unseen connections.

What?

What about "distillation"? What does that mean? A distillation is the essence of a book – what it is really about. A distillation is not a summary, not simply a shorter version of the original, not CliffsNotes by another name, not plagiarism. It is a rewritten, reorganized version that reveals themes the original author might have buried under important but obfuscating detail. The distillation helps people understand the context and structure of information.

THEMES

Various interrelated themes tie the books together. All relate to major changes in civilization as we know it (etc). Over the past year I've tried various themes as if they were new suits. It's been tough. But these last ones seem to fit.

Stuff and Desire

A major theme is stuff (which can be material or immaterial), desire (resulting from need or greed), and the collapse that happens when there is a mismatch. This theme is fairly obvious in the first four distilled books...

- The Great Disruption – Paul Gilding
- Upside of Down – Thomas Homer-Dixon
- Coming Apart – Charles Murray
- Collapse – Jared Diamond

Black Swans

Another theme, less obvious, but still important derives from Nassim Taleb's notion of Black Swans. These are unpredictable events that can have surprising results. Black Swans that suddenly increase demand or decrease the supply of stuff could trigger a collapse. (Conversely, Black Swans that decrease demand and increase stuff could forestall a collapse.)

Flat Earth

Thomas Friedman wrote Flat Earth. His related theme is that a connected Earth is a flat Earth. A flat earth allows events in one place to be rapidly transmitted to another place. Those events can include collapses.

ASYMPTOTES

Asymptotes are my theme.

The authors never mention asymptotes. However asymptotes help visualize the relationship between stuff and desire and what happens when stuff runs out.

 (Mathematicians beware; the following is heavily metaphoric. All be aware, the following is the work of an educated redneck. The observations seem obvious, but who knows for sure.)

What Are Asymptotes?

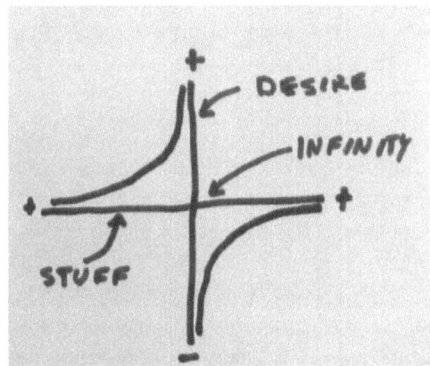

Gilding quotes a professor named Tim Jackson....

"The global economy is almost five times the size it was half a century ago. If it continues to grow at the same rate the economy will be 80 times that size by the year 2100."

Gilding's point is that such expansion cannot continue. It will hit the wall - several walls - of finite reality. Gilding cited a study which says that we are presently consuming resources at the rate of 1.4 planets. The underlying issues are the loss of global biodiversity and changes to the global ecosystem. But the one that will grab us (by the throat or crotch - pick your sensitivity) is the end of growth.

Stuff, in quantity and quality, defines us. It is how we judge ourselves and other people - the cars we drive, the clothes we wear, where we eat, where we shop. Growth is key to economic policy. An economy that does not provide growth is considered to be failed. Business is predicated on ever-increasing demand. Expanding populations require growth to create new jobs for new entrants into the workforce. But the demands of efficiency push automation which reduces the available jobs.

It is like a giant, system-wide Ponzi scheme that is about to collapse; the future on which such schemes are based is about to arrive. It will probably go down in a nonlinear fashion (maybe precipitated by a "Black Swan" event - maybe by the accumulation of lesser events) because that's how the curve goes - slow on the way up until near the end, when it goes almost straight up, then fast on the way down finally flattening with random stutters - not a neat bell shape at all.

The other point Gilding makes is that none of this will be easy. He says that we are essentially addicted to growth and that like any addict we will not willingly give up our drug. We will lie, fight, and deny the reality in front of our noses. We will have to hit the walls of finite reality several times, really mess ourselves up before the truth penetrates our bloody, battered, dim-witted heads.

WE TRY TO BUY HAPPINESS WITH STUFF
A British study indicates that the "loneliness index" goes up with increased wealth.

For our purposes, an asymptote is two straight lines that cross, two curves that approach but never touch the straight lines and a center. Here is what the parts mean:

- The horizontal line represents stuff (material or immaterial).
- The vertical line represents the desire for stuff (it can represent time).
- The curve represents the relationship between desire and stuff. Starting out, when the supply of stuff is great relative to desire the curve has a gradual slope. As stuff starts to run out and desire increases the slope starts to get steep. As stated below even if desire remains constant, the passage of time can drive the curve up.
- The center represents infinity. It is what gets magically crossed

When the curve gets too steep it wobbles and falls. Infinity is crossed. (Taleb might say that a Black Swan happens.) If the fall is not too great, the curve might re-generate itself - desire decreasing and stuff increasing until the whole thing starts over again. This is shown in the lower right part of the curve.

Desire and Time

Desire does not have to increase for there to be a collapse; desire could even fall. What counts is that the supply of stuff runs out. The asymptotic curve could climb by plotting the diminishing supply of stuff against the passage of time. Mostly in this book we'll pretend it is desire and not time that causes stuff to collapse. Otherwise we'd have to bring up entropy and the arrow of time which is almost always an underlying theme – but can be even more confusing than the rest of it.

The Curves Have Names

The first three books are explicitly about collapses. (Diamond's book, Collapse, is different.) I have given names to the asymptotic curves related to these collapses. The names are based on the kind of stuff being desired.

In Gilding's *The Great Disruption*, natural resources are being depleted. The name given to the curve representing the relationship between diminishing stuff and desire is The Consumption Curve.

In Homer-Dixon's *Upside of Down* the stuff being desired is energy (although other stuff is mentioned – it's a complicated book). As the desire for energy pursues ever-diminishing supplies, systems compensate by becoming more and more complex. Homer-Dixon says that at some point, systems will become so complex they fall apart. The name given to the curve representing this relationship is The Complexity Curve.

In Murray's book *Coming Apart*, jobs are being consumed and becoming less available – specifically the jobs of the lower classes. As jobs become less available happiness decreases and unhappiness increases – and a variety of other conditions. The name given to this curve is The Unhappiness Curve.

STOPPING COLLAPSES?
Controlling or stopping collapses is a matter of concern to Gilding, Homer-Dixon and Murray. (Diamond mostly describes collapses that have already happened – Tableb and Friedman don't deal directly with collapses.)

Collapses happen in finite systems.
Collapses happen when desire (or time) continues but stuff runs out. In the examples described in Diamond's book Collapse, collapses happened in regions isolated by geography – on islands surrounded by an ocean or in regions of productive land surrounded by regions of non-productive land. These are finite systems. As described in Thomas Friedman's Flat Earth, our entire planet is now a finite system. We range the planet using up stuff wherever we go. Supply networks stretch. Everything becomes connected to everything else. A collapse in one place triggers a collapse everywhere.

Two Ways To Stop/Defer Collapses
There are two ways (obvious it seems to me) to eliminate or at least defer a collapse – decrease desire or increase the supply of whatever is desired.

Increase Supply of Stuff
When the problem begins to be noticed and pain felt, the supply side of the equation is often (in my experience) addressed first. More stuff is obtained locally, dipping into diminishing resources. When that

starts to run out, supply is obtained from other regions. More and more efficient systems are devised. As the cost and availability of the preferred supply increases, alternative stuff is used. For example fracking is used to obtain hard-to-get oil and gas. Tilapia replaces tuna when tuna is overfished and becomes too expensive.

Viewed asymptotically, the curves (consumption, complexity, and unhappiness) get steeper and steeper. At some point a curve can no longer be supported. There is a collapse.

Decrease Desire

Desire, if stemming from the *need* for stuff in order to live, cannot be decreased unless the population is maintained or reduced. And even then a reduced population could make up for reduced need by increasing desire. (This has already started to happen in some advanced countries.)

If stemming from *greed*, desire can be decreased if the habits of the population are changed. Generally people resist this solution.

Greed can lead to need which can lead to collapse when stuff diminished.

In Collapse, Diamond describes three societies, New Guinea, Tikopia, and Japan, where desire was managed and collapses prevented. This was done by maintain populations and controlling demand.

ASYPMTOTC RELATIONSHIPS IN DISTILLED BOOKS

All the books distilled in Crossing Infinity deal in one way or another, perceptibly or imperceptibly with the concept of the asymptote and how it topples. (Taleb would call the gently sloping part of an asymptote Medicocrtistan and the steep part Extremistation – the place where Black Swans can trigger a collapse.)

Following are summaries of the books from asymptotic points-of-view.

The Great Disruption

The curve is consumption. When consumption becomes too great things fall apart.

In Gilding's book, desire results from a combination of need and greed. Desire and stuff create the consumption curve. The Great

Disruption described by Gilding is a wobble in the consumption curve, a lesser disruption that foretells the Great Collapse.

The Upside of Down

The curve is complexity. When complexity becomes too great, things fall apart.

The desire focused on by Homer-Dixon is the pursuit of the energy needed to fuel growth. Supply is the source of available energy. The ever-increasing search for new sources of energy results in ever-increasing complexity of the systems used in the search – which according to Homer-Dixon makes the systems more fragile and vulnerable to a fall. Eventually, when demand exceeds supply and systems become unmanageable, the complexity curve starts to wobble, and a Pulse happens (what Gilding calls a Disruption). If we pay attention to the pulse and control demand, the ultimate collapse can be avoided.

Coming Apart

The curve is unhappiness. When unhappiness gets too great there might be a revolution. This is the least obvious curve of all.

Murray's notion of a collapse is when society comes apart as a result of increasingly large lower and upper classes pulling population from a shrinking middle class. Murray says that this movement is caused by changing class structure which in turn is caused by a loss of responsibility, freedom, and a diminished adherence to what he calls the founding father virtues. These changes are concentrated in the lower class but can spill over into other classes. That will result in a failure of leadership and a loss of values across all classes.

Of underlying importance to Murray is how these factors affect the happiness of people that comprise a class.

One of the major factors contributing to happiness is jobs.

Consider the lower class. There is an increasing demand for the ever-diminishing manual jobs that can be performed by the ever-increasing lower class. Supply is represented by availability of these jobs. People in the underclass are not happy.

Contrast the plight of the lower classes with the condition of the New upper classes. They tend toward "knowledge work" of which there is an increasing supply. This increase is due in part to the increase in complexity described in Upside of Down. These people are generally happy.

The supply (of jobs) and demand (by job seekers) can be pictured in an asymptotic manner to create an unhappiness curve.

The asymptotic collapse happens when the discontent of the lower classes reaches a threshold. This he says will start in Europe when that model of the welfare state fails. In asymptotic terms, the European collapse will provide the shudder in the happiness curve needed to wake us up to the general collapse to come.

Collapse

Collapse is about infinities that have already been crossed.

Jared Diamond discusses societies, ancient and modern, that have collapsed. All the collapses occurred despite efforts to increase supply. He also discusses a few societies that controlled demand and did not collapse.

Black Swan

Taleb (without meaning to) describes how asymptotic curves fall and infinities get crossed.

Black Swans, positive or negative unanticipated events, can result in sudden changes in the asymptotic curves that describe supply and demand relationships between processes.

Negative Black Swans can push an asymptotic curve nearly vertical, causing infinity to be crossed. For example a curve representing consumption could suddenly be pushed toward the vertical if a Black Swan caused either a sudden increase in demand or a sudden loss of supply.

Positive Black Swans can cause the curve to level out For example a Black Swan that caused a sudden drop in demand or a sudden increase in supply could cause the curve to become less steep, perhaps even to slope downward.

Flat Earth

Friedman discusses how events – including collapses - in one place can spread to another place.

The globally connected flat earth described by Friedman is a means by which an asymptotic feedback loop can be happen. This affects this which affects that which affects that which affects this. Demand affects demand which affects demand etc. As the loop proceeds the curve gets steeper.

.

.

THE GREAT DISRUPTION

Bloomsbury Press 2011

ISBN 978-1-60819-223-6

Paul Gilding

The curve is consumption.

In this book, demand is the unlimited desire for a finite supply of stuff (again, material or immaterial). Demand and supply create the consumption curve, going up and falling down. The Great Disruption described by Gilding is a wobble in the consumption curve, a lesser disruption that foretells the Great Collapse.

SUMMARY

Gilding describes what is happening, what will happen. His is a book of Greats: Great Disruption, Great Awakening, and Great Collapse.

A Great Disruption is about to happen. It will be the result of a millennia of growth that accelerated in the last century. Things will fall apart.

Gilding says there will also be a Great Awakening when people realize what is going on. If we are lucky the Awakening will precede the Disruption. We can't stop either; there isn't time. But if we anticipate the Disruption and act correctly, things could ultimately work out. The next millennia could be one of peace and understanding.

If we don't act correctly there could be a Great Collapse. We could witness a Mad Max/Cormac McCarthy style dystopia.

SOMETHING BIG IS COMING

We are about to be seriously disrupted. It won't be a once-in-a lifetime, or even once-in-a-century event, but something that happens once-in-a-civilization, something that defines an era —like the Renaissance, Enlightenment, or Dark Ages.

(Gilding says it might be even bigger, redefining us at the evolutionary level. People with certain propensities might breed more. Those that survive might think and behave differently.)

Disruptions have already occurred in various third world countries - Haiti, Bangladesh, Rwanda, and others. Much of Africa has been disrupted. Populations have starved, died of disease, gone without fresh water. People have suffered the upheavals that Gilding says will happen to first-world countries unless we react properly and don't let the disruption slide too far.

At the time of this writing, first world countries are still in reasonably good shape. Although financial and political discord is common not many people are actually starving. Nobody is fomenting revolution at the time of this writing – not in the U.S. anyway. Quite a few people are getting rich.

So, why are disruptions looming?

Principally because the earth is running out of room and resources. There are too many people consuming too much stuff. The planet cannot replenish itself. It is like a global Ponzi scheme in which current prosperity is based on the anticipation of future gain. We are robbing Peter to pay Paul. We have leveraged today into tomorrow and it is just a matter of time before the future arrives and we discover that our gains are gone.

Although most people don't know an end is coming, many people seem uneasy. They join fringe political groups and are attracted to tales of gloom and doom. They suspect something and will not be surprised when their suspicions are confirmed.

Gilding calls this moment of awareness the Great Awakening. It might be sudden or gradual. It might occur before, during, or after the Great Disruption.

The Great Awakening (upper case) might be triggered by an ecological/economic Pearl Harbor. For example, there might be a pandemic resulting from the spread of pathogens from third to first world countries. Or the accelerated melting of artic ice, causing coastal regions to flood might result in an Awakening. Certainly there would be increased awareness if the Gulf Stream failed, paradoxically triggering a new ice age in the midst of global warming.

Alternatively an awakening (lower case) might be a tipping point of public awareness that results from a gradual accumulation of lesser events. People might just wake up one day and see what they have not seen before.

In asymptotic terms the consumption curve stutters a little bit or a lot.

Whatever the proximate cause, Gilding claims that by the time the awakening (great or otherwise) happens it will too late to change course. Nothing can help then. The Great Disruption will happen (or will have already happen).

However, and this is the point of his book, we can stop ourselves from falling completely over the edge into a Mad Max/Cormac McCarthy dystopia - especially if we are lucky enough to have a preceding awakening . If we put ourselves on a wartime footing the human species might not fall completely off the edge. We might stabilize the Fall, fix the climate crisis and kick our addiction to growth – which caused the problem to begin with. We might avoid The Great Collapse.

THE HIPPIES SAW IT FIRST
Flower children were the first to perceive the threat and their children will lead the movement to do something about it.

The environmental movement has been going on since about 1962 when Rachel Carson published *Silent Spring*. Gilding imagines the voices of the generation that followed as a collective scream – like the one emanating from the face in Edmund Munch's painting, The Scream.

Following is a summary of major events taking place during the Scream period...

- <u>1959</u> – U.S. government announced dangerous levels of a weed killer in cranberries from Washington and Oregon.

- <u>1962</u> - Rachel Carson published *Silent Spring* exposing dangers of pesticide. She was vilified by the chemical industry but her book became a best seller and led to the development of the EPA.

- <u>1969</u> – The Cuyahoga River in Cleveland burned.

- <u>1971</u> – Greenpeace was founded.

- <u>1972</u> – The United Nations Conference on the Human Environment was held in Stockholm.

- <u>1972</u> – The *Limits to Growth* was commissioned by the Club of Rome and produced by experts in system dynamics at MIT. The book documented 12 possible futures examined by the computer model World3. All scenarios tested interactions between nonlinear (ever increasing) growth in a planet of finite resources. Some scenarios were deliberately farfetched, included to give rigor to work. The report concluded that given verified human impact on the planet, environmental and economic collapse was inevitable. Industry critics and conservative politicians focused on the farfetched scenarios and condemned the book. However, according to Gilding the results have come to be generally accepted (by whom one wonders). He also notes that predictions of the model have proven to be surprisingly accurate.

- <u>1984</u> – 15,000 people were killed when Union Carbide plant in Bhopal India released tons of toxic gas.

- <u>1985</u> – French intelligence agents operating with the approval of French government bombed the Greenpeace vessel Rainbow Warrior.

- <u>1987</u> – Governments around world (including the conservative governments of Margaret Thatcher and Ronald Reagan) agreed to phase out chlorofluorocarbons (CFCs) – which were creating a hole in earth's ozone layer (thereby increasing risk of cancer).

- <u>1997</u> – the Kyoto Protocol was approved and Earth Summit+5 was held.

- <u>2002</u> – Earth Summit 2002 was held.

During the Scream period there was an increased awareness of environmental issues. Some political leaders started to talk the talk.

And as the above list shows there has been some action. However the action has been inadequate and the problems have gotten worse.

Gilding quotes a speech given by Winston Churchill in 1936 in which he says that the time for drift is past. "... we have entered a period of danger... we are entering a period of `consequences ... we cannot avoid this period, we are in it now..."

MOST EVERYBODY ELSE HAD TROUBLE SEEING IT

We are biologically programmed to respond to immediate threats – only the prescient few – old hippies, their progeny, some scientists, some politician, perhaps English majors can see beyond the surface obvious to a deeper obvious.

Guiding observes that people have trouble understanding the dire forecasts of global ecosystem studies because things look pretty good right now - at least in the developed world. We are not born with the mental equipment to understand anything not happening in front of our noses. Such insight has to be acquired. Guiding says that if we don't accept the science (which is about as unanimous as science gets) and wait for evidence that we can see, it will be too late. The Great Disruption/Collapse will be on us. It almost is.

NEVERTHELESS, IT WILL HAPPEN

Even if old hippies and their progeny see it, the changes required at this time are too great - the inertia of systems against change cannot be practically overcome.

Changing economic/environmental practices has always been framed as a choice. If we change this thing we will avoid that thing - causing something else to happen. We always had a choice. Not now. He says, "This means any hope that we can mobilize the massive intervention required to avert the crisis is a false hope. In combination the evidence all points to one conclusion. We cannot now avoid the crisis of the Great Disruption."

OUR LOVE OF STUFF GOT US HERE

We are wedded to growth and its paired drivers, consumerism and stuff.

Other studies have shown that after basic needs are met (about $15K per capita at the time I started this distillation) increased wealth does not buy happiness.

Founders of economics did not propose continual growth. John Stuart Mill said that a "stationary state of capital and wealth... implies no stationary state of human improvement". John Maynard Keynes thought that the "economic problem" would be solved and that society would then "prefer to devote further energies to non-economic purposes".

Homer-Dixon Asides – Stuff Does Not Give Meaning to Life

Quoting Homer-Dixon's *Upside of Down* (another book on the same topic distilled in Crossing infinity)...

"Despite the fact that our lives are saturated with stuff, that we've already reached a level of material abundance unimaginable to previous generations, and that more money and possessions add little to our happiness, we must be made (by business) to feel chronically discontented with our lot."

"Our economic role in this culture of consumerism is to be little more than walking appetites that serve the function of maintaining our economy's throughput."

"Our psychological state is comparable to that of drug addicts needing a fix: buying things doesn't really make us happy except perhaps for the moment after the purchase. But we do it over and over anyway."

"Why? There are many reasons. But the central and often overlooked one, I think, is that consumerism helps anesthetize us to the dread of empty lives - lives that modern capitalism and consumerism have themselves helped empty of meaning."

THINGS STARTED TO HAPPEN IN 2008

Gilding argues that 2008 was the year The Great Disruption started to be seriously felt. These two "crash" indicators occurred:

- *Ecosystem hit tripping points*
- *Resource limitations forced prices up*

Ecosystem Tripping Points

Nonlinear, self-reinforcing events happened (*the consumption started to get really steep, maybe tremble a little*):

- <u>Melting of northern icecaps</u> accelerated, exposing dark blue ocean which heats faster which accelerates ice melting.
- <u>Melting of frozen tundra</u> accelerated the release of large quantities of methane which is a greenhouse gas which traps heat which accelerates release of methane.
- <u>Ocean acidification</u> increased which (1) reduces the ocean's ability to absorb CO_2 (2) prevents shellfish from forming shells and (3) coral reefs from growing and heating the atmosphere.

Prices Increased

Increasing prices indicate that various "walls" have been hit. For example:

- Oil became increasing difficult to extract - easily accessible oil has all been found and is being used - meaning "peak oil" has or soon will happen. As a result oil prices went up.
- Global food prices increased because...
  a. There are more people demanding more; there is less arable land (due to development, over-pumped aquifers, falling water tables, over-allocated rivers, diminishing crop yields, expanding deserts, etc.)
  b. Rich people (us – compared to most of the world) eat more and better food.
  c. Corn is diverted to biofuel instead of food.

Leaders Started To Take Notice

The reaction to Gilding's message changed in 2008:

- When he first started presenting his story to social and business leaders Gilding was regarded as "intellectual entertainment" - written off as an "extremist and merchant of doom". Gilding believes that these leaders are good people who have invested their professional and personal lives in the notion of growth. It became a given, a fact of life not to be challenged.

- After the crash of 2008 was well under way, Gilding got different reactions. The same leaders felt that something was going on – a change had taken place. Many agreed in principle with Gilding - in the direction that things were heading, but were more optimistic that something could be done to forestall the ultimate collapse.

Thomas Friedman Also Points To 2008

Gilding quotes Thomas Friedman...

"What if the crisis of 2008 represents something more fundamental than a deep recession. What if it's telling us that the whole growth model we created over the last 50 years is simply unsustainable economically and ecologically and that 2008 was when we hit the wall - when Mother Nature and the market both said, 'No more.' "

The Military Agree

Military people already warn about the risks of collapse...

- Retired Marine Corps General Anthony Zinni said, "The 2007 report concluded that climate change would act as a threat multiplier by exacerbating conflict over resources, especially because of declining food production, border and mass migration tensions, and so on - increasing political instability and creating failed states - if no action was taken to reduce impacts."
- Thirty three retired generals and admirals wrote in April 2010 report to the Senate, "climate change is threatening American security... it exacerbates existing problems by decreasing stability, increasing conflict, and incubating the socioeconomic conditions that foster terrorist recruitment. The State Department, the National Intelligence Council, and the CIA all agree, and are planning for future climate-based threats."
- A secret 2004 Pentagon report noted, "Disruption and conflict will be endemic features of life... once again, warfare would define human life."

THE END OF THE BEGINNING
Our complex, growth-addicted culture cannot survive as is.

Although details of the future can't be predicted, Gilding thinks the outline is clear. The end might have begun in 2008 or it might not happen until the next decade. But it will happen. Growth will stutter to a stop. After that civilization will either...

- Stabilize and evolve to a higher plane (Gilding's belief).
- Or become much simpler (e.g., it will collapse like the cultures described in Jared Diamond's book *Collapse*).

The fall of a system (ecological/economic) resembles the end of a human life. Gilding's friend Dr. John Collee describes it this way...

"Every patient with an incurable illness will ask how long they have to live. The answer goes something like this: 'No one can say how long you may live, because every individual is different, but focus on the changes you can observe and be guided by those. When things start changing for the worse, expect those changes to accelerate. So the changes that have occurred over a year may advance by the same degree in a few months, then in weeks. And that is how you can judge when the end is coming.'"

"Planet Earth, being a web of complex self-regulating systems, operates very much like a human body. Terminal illness gives us the template for most forms of ecological collapse. One set of changes initiates another, and so on in a downward cascade of negative feedback until the whole system falls apart."

Like the terminal patient we must look for certain signs. Gilding says to look for an accelerating cascade of...

- Ecological, social and economic shocks driven by climate change.
- Increases in food prices due to demand and lower output.
- Diminished water supplies, fisheries, and agricultural output resulting from damaged ecosystems - further increasing prices.
- Increased oil costs as peak oil happens (if it has not already happened).

- <u>Falling stock markets</u> driven by fear and uncertainty.

Does all this mean the world is coming to an end? Gilding says not necessarily - but it does mean we (or the children and grandchildren of those of us presently approaching senility) are in for a ride. The Beginning of The End

Gilding says change is coming and we'd better be prepared and resilient. This is true for individuals and institutions - private and public.

He notes four aspects of change to be especially aware of...

Impacts of Climate Change on Security and Economy

There will be food shortages, supply shocks, price volatility - regardless of our response. Contributing factors include:

- <u>Industrial agriculture</u>. It depends on nitrogen fixation which depends on carbon which is nonrenewable. It can run out.
- <u>Integrated, just-in-time food chains</u>. Although efficient - delivering low-cost food across the globe, such systems are also vulnerable. Transportation failures, terrorist attacks, etc. could leave some ends of chain with only four days of food on shelves.
- <u>Increased competition</u> for food as countries like China and India become wealthier.
- <u>Less land</u> available for farming.
- <u>Competition</u> between crops gown for food and for biofuel and other industrial uses.

Shortages and disruptions will create unrest and instability, especially in poorer countries.

Shifts in Economic Competitiveness

There will be winners and losers as The Great Disruption occurs and countries put their "war plans" in effect. If what Gilding says will happen, the oil producing countries and the oil industry will be losers. So will the coal industry. Generally all those who deliver carbon based energy will suffer. However, those who develop and provide energy from renewable resources (solar, wind, hydro, etc.) will prosper.

Gilding sees the most interesting competition taking place between the U.S. and China. The Great Disruption is closer to happening in China due to the greater environmental damage that has already taken place. Consequently China is closer to implementing its own war plans. It is hitting the "physical limits of its economic growth model". Gilding quotes Tom Friedman (the Flat Earth guy):

"Yes, China's leaders have decided to grow green - out of necessity because too many of their people can't breathe, can't swim, can't farm and can't drink thanks to pollution from its coal- and oil-based manufacturing growth engine. And, therefore, unless China powers its development with cleaner energy systems, and more knowledge-intensive business without smokestacks, China will die of its own development."

China is seriously pursing a low-carbon economy. So are India, Brazil and South Korea.

Gilding thinks China might win the competition with the US because it is not wedded to a market-based economy. Further, its government, less burdened by the need to observe Western democratic freedoms, might be more efficient. It might respond faster to the problems. (Of course China has a history of mistakes – such as The Great Leap Forward.)

Loss of Moral Authority

Countries, systems and economic models that win the economic war started by The Great Disruption will have increased moral authority over those who don't. For the latter two thirds of 20th century that war was won by the West - by the US in particular. This might not be true in the 21st century.

Reaction of "Victims"

Some small low-lying countries and some entire regions will cease to exist - disappearing under water or sand. Even if the causes of global warming were ended today, the effects have momentum and will continue for decades. Gilding sees reparations being sought by individuals, groups, regions, countries. Some redress will be pursued in courts of law. Some in the streets.

HOPE ANYONE? (SURVIVING THE GREAT DISRUPTION)
Gilding turns a corner, moving from despair to hope. But it is not an easy move to make.

To survive The Great Disruption our attitudes have to change. According to Gilding, we must...

- Accept that things are going to get really bad. People are going to suffer. People will die. We have to prepare ourselves - physically, economically, and psychologically. It will be like a world war.
- Drop our old ideas about how change occurs - understand that change generally does not happen slowly, in a planned manner, but fast - in this case, as a series of really bad Black Swans.
- Evolve a new set of values, politics, and personal expectations. We have to shed our ethic of consumerism. (See next asides below for Homer-Dixon's views on values.)
- Accept the idea that we are not saving the planet, only ourselves - our species. Although we might wipe out 50% of present day biodiversity, in 100 million years (a moment in the life of a planet) Earth will have long moved past us. It will get along just fine.

Gilding says that despair is a legitimate reaction to what is going on. He notes that some experts in this field - including James Lovelock, author of *The Vanishing Face of Gaia: A Final Warning* and Clive Hamilton who wrote *Requiem for a Species* - believe we are not smart enough to come back from the Fall, that it is too late.

Gilding does not agree. He believes that the despair and hopelessness evidenced by the above writers (and until 2008 by him) is step two in a three step process...

1. The first step is denial which itself has two sub steps. Initially we don't believe what is happening. Then, we enter what Gilding calls "denial breakdown". We more or less believe the science, acknowledge that it makes sense but do not believe the full implications. Or, we accept the implications but do not yet feel the impact - do not move from intellectual

understanding to emotional understanding. (Gilding says that climate deniers and anti-science skeptics can be ignored; they will be overcome by events.)

2. The second step is <u>despair</u>. We begin to sense what is really going on. We see the species lost - the animals starving, dying – the children starving, dying. We see a generation of underemployed, drifting young people (our children, grandchildren?) - everyone a potential Mad Max - and us the wandering man in Cormac McCarthy's *The Road*.

3. The final step is <u>acceptance</u>, what Gilding calls the awakening (term "Great Awakening" was coined by Professor Jorgen Randers). This happens when we move past grief - when we start to do something - when we understand that many of the drifting young people mentioned above can be productively engaged in saving our species. Stuck as we are between denial and despair there is nothing practically we can do to prevent The Great Disruption. But we can change what happens on the other side - whether we collapse, as described in Jared Diamond's *Collapse*, or whether we transcend.

Gilding says that the trick to moving past grief into acceptance and action is the belief that we can actually solve our problem (again, not prevent the Disruption but manage what follows). Although it might seem that we are doomed, Gilding points to our history of waiting until last moment to act in the face of imminent disaster.

He uses WWII as an example. Although there were clear indications in 1933 that Hitler was a threat to world survival, people in the US and Britain remained in denial and despair until the last minute. The crisis was well underway before leaders like depressed Winston Churchill, grandiose Franklin Roosevelt, and monstrous Joseph Stalin finally rallied their people into a level of action unimaginable before the war. We did not prevent the war but we did save civilization from Hitler. Gilding sees that same level of action happening in response to The Great Disruption. We can't stop the war but we can win it

Gilding says that a single event might be viewed (at least in retrospect) as the ecological/economic Pearl Harbor that calls us to action (an ad hoc black swan – the Great Awakening).

Homer-Dixon Talks about Survival

Paraphrasing Homer-Dixon on values in *The Upside of Down*...

- We need to move beyond strictly utilitarian values which only express our likes, dislikes. This gives rise to our consumer oriented culture. Consumed by consumerism, we become less resilient, more vulnerable to unexpected nonlinear events.
- We also need to move beyond the unthinking, politically oriented notions of fairness, and right and wrong which often pass for spiritual belief. This too leaves us rigid, vulnerable to nonlinear events.
- We need to move into the realm of spiritual and existential values that are "compatible with the exigencies of the natural world".
- According to Homer-Dixon, such values recognize that:
  1. Energy and the laws of thermodynamics play a key role in our survival.
  2. Certain kinds of connectivity are dangerous.
  3. Many natural systems (including all adaptive systems) behave in a nonlinear manner. In the words of Nassim Taleb we live in Extremistan.
- We need to move from a growth imperative to a resilience imperative.

Signs and portents seen late at night on CNN while I was writing the first draft of this book...

- Drought is killing people in the Horn of Africa. Thousands of people are on the move. Last night I had to turn away from a picture of starving child held by its mother. The mother was dressed in a dirty robe that had once been colorful. The child was a skeleton covered in skin, a delicate little drum, its ribs a marimba. Although it certainly doesn't matter to the mother and child if the drought is the result of climate change, maybe it is. Maybe an obscene black swan waddles among the corpses.
- Big Oil is digging oil sand in the Canadian wilderness. Tremendous amounts of energy are needed to extract the oil from the sand. Tremendous quantities of greenhouse polluting gases are being released. The landscape looks like

the surface of the moon. Maybe Homer-Dixon is right. Maybe Peak Oil has happened and we are now forced to bizarre and extreme solutions in our search for energy.

- Jabbering crowds throw bodies off a bridge in Syria - the bloody corpses whirling gaily through the air to flop in the water beside their dead companions. The Arab Spring moves into late Summer. I once read or heard somebody say that rising food prices had something to do with this.

PREVENTING THE GREAT COLLAPSE

Gilding states his belief that we have the power to prevent The Great Disruption from becoming The Great Collapse.

He believes that in the near future, an ecological/economic Pearl Harbor event will break us out of denial and despair, move us to the Great Awakening. (Or an accumulation of lesser events which result in a tipping point of awareness leading to a lower case awakening. As noted before he does not seem clear on this.)

The implication is that the black swan event (or the accumulation of lesser events) must be bad enough to convince even anti-government conservatives that action is necessary. The awakening will require death, destruction and economic disruption. But the event/events must stop short of triggering ecological feed-back loops that result in a total climate collapse.

In other words we will have to be lucky - if you can call a Pearl Harbor style event (or the conditions that result in a tipping point of awareness) luck. That is what Gilding requires you to believe.

So, in this spirit of belief...

After the event occurs (or the series of lesser events – perhaps symbolized by one dramatic Pearl Harbor event) the "dam of denial" will break. We will move to a war-time footing.

Action will take place on two fronts.

1. The old economy and systems will react first. Driven by a vested interest in maintaining established ways, yet aware now of the extent of the problem, this part of our society will fight the initial tactical battles. Spending great sums of

money, expanding the power of government, the old order will attack the problem of greenhouse gas emission, or whatever it is that pushes us to the edge. Such actions might even create a period of financial growth.

Gilding believes that in this war-like atmosphere, almost anything can be accomplished - at least in the short term. However, he thinks that the old order will not willingly give up the idea of continued growth. The old order will accept that continued material consumption cannot go on, but will believe (as it believes now) that material growth itself can continue - maintained by new technologies and efficiencies. The old order will argue (as it argues now) that we are basically a greedy, competitive species and that this style of economy suits us best.

2.   This is where the second front of the war comes in. Knowing, as Gilding knows, that the laws of physics do not allow unlimited material growth in a finite world, some people will coalesce around a new economy and style of living. This new order will fight the strategic long term battles - over a period of 40 or so years. It will move us past consumerism and stuff.

AFTER THE DISRUPTION – PLAN OF ACTION

Accepting the belief (the hope?) that a Pearl Harbor level ecological/economic event - or an accumulation of lesser events - will cause an awakening; the next step is to come up with an action plan.

Gilding and his friend Professor Jorgen Randers, one of the authors of the Club of Rome's *The Limits to Growth* (and recently *The Limits to Growth: The 30 Year Update*) propose this plan for the next 100 years.

The first action - likely undertaken with or without a plan - will be to reduce greenhouse gases. Climate warming will probably be seen as the cause of the Pearl Harbor style ecological/economic event (or the accumulation of lesser events). Given that the definition of this event (or events) includes the capacity to rouse the old order to action, it therefore follows that the old order will be roused and will act.

Since greenhouse gases are responsible for global warming the question arises: how much warming can be allowed? What should be the goal? The current consensus is that a two degrees centigrade rise over pre-industrial revolution levels can be allowed - not because it is "safe" but because it is the best we can do politically.

Gilding disagrees. Quoting Winston Churchill, Gilding says we must do "what is necessary" - not just what is politically expedient. A two degree rise is likely to produce great damage and disruption. A one degree cap is safer. Gilding thinks this will be politically feasible given the impact of The Great Awakening (that results from The Great Disruption).

CONDUCT A ONE HUNDRED YEAR WAR

Estimates are that the awakening (Great or otherwise) will occur sometime around 2018 and that it will (must) be followed by a long term period of action

Gilding says that the following actions need to occur in the time frames indicated.

Only the strong shall survive.

Years 1 -5 (2018 - 2023)

Reduce greenhouse gas emissions by 50%. Not only will this begin to clean up the atmosphere and start the cooling process it will shock our political and social systems into action. Following are some specific goals recommended for the first five year period:

- Cut deforestation and other logging by 50%.
- Close 1,000 dirty coal power plants.
- Retrofit 1,000 other coal power plants with carbon capture and storage.
- Ration electricity and increase gas mileage.
- Erect a wind turbine or solar plant in every town (if only for psychological boost).
- Create huge wind and solar farms in suitable location.
- Ration use of dirty cars to cut transport emissions by 50%.
- Strand half of world's aircraft.
- Capture or burn methane.
- Move away from climate unfriendly protein.
- Bind one gigaton of CO2 in soil.
- ua government and community-led "shop less live more" campaign (hard to imagine old order sanctioning this).

Years 5 - 20 (2023 - 2038)

Move world to net zero climate emissions. Requires technological and social innovation.

Years 20 - 100 (2038 - 2118)

Achieve a stable climate and a sustainable economy. Reverse emission; remove CO_2 from atmosphere.

Obviously all this seems impossible. Gilding says to look back at the changes brought about WWII. (One change struck me. Four days after Pearl Harbor the government ordered a halt to the production of private vehicles.)

GET STARTED WITH WAR

If you believe that The Great Disruption will really happen and that a war plan is necessary then you should do something – you should get started with the war.

Business Responds to Changes in Environment

According to the Austrian economist Joseph Schumpeter, markets are engines of creative destruction - "a process of industrial mutation that incessantly revolutionizes the economic structure from within, incessantly destroying the old one, incessantly creating a new one."

Although Gilding compares The Market (my emphasis) to a rain forest he doesn't extend the metaphor to explicitly say that markets are instruments of economic natural selection. Therefore, making the analogy for him - companies, like living organisms, change in response to changes in the environments in which they function.

However...

In natural systems changes are due to random DNA mutations - which either provide advantage or disadvantage in constantly changing environments. If an advantage is forthcoming the organism survives to pass on its DNA; otherwise the organism fails and becomes an evolutionary dead end.

In economic systems changes are made more-or-less consciously by business people in response to changes in The Market - the business environment. When changes work, the business makes money - it prospers. When changes don't work, the business loses money - it fails.

Change or Die

Given Point Two, the way to change business and fight the one-degree war is to change the environment in which business operates. Change The Market and at least some businesses will implement the changes necessary to make money. Businesses that adapt will prosper, those that don't will fail.

In the case of the one-degree war, the aspect of The Market most subject to change and control is the regulatory environment. Change the regulatory environment and business will follow; it has no choice - like any organism, living or otherwise, business has to play by the rules of the environment.

Of course the present national environment of which the business environment is a part, is decidedly anti-government, anti-regulation. That will not likely change unless The Great Disruption happens.

GILDING SAYS IT WILL HAPPEN BECAUSE IT MUST

We cannot let The Great Disruption become The Great Collapse.

Gilding attempts to answer those who ask, "How will all this happen? There is so much cheap coal in the ground, so much natural gas - even petroleum - how could people simply give that up?"

His answer is that he doesn't know exactly how it will happen. But he is certain that it will happen because it must –a Great Awakening that anticipates the Great Disruption or that follows on its heels . He believes there will come a time in the not-to-distant future when it becomes obvious that we cannot continue down this same road - that something must be done to stop the slide past a one or two degree C rise in temperature over pre-industrial levels.

He says, "... it is clear governments have to act, so they will. That action must result in dramatic reductions in CO_2 emissions, and the science says that must result in the decline of coal and oil, followed soon after by gas. There are no realistic scenarios where this can be achieved if we wait past 2015-2020, unless we decide to go past two degrees (centigrade) of warming. Given that all the world's major governments have agreed not to, the logic of the economic risk flows pretty easily."

Gilding says the future will be characterized by "dramatic and discontinuous change" (implying nonlinear events - Black Swans) - whether we go over the edge or just get close. The difference seems to be the kind of planning we can do. If we go over the edge and "the economy collapses under the weight of climate and sustainability impacts" no planning will be possible - the level of chaos will be too great.

However, given that the latter scenario means things will pretty much go to hell anyway, Gilding doesn't dwell on it. Instead, he assumes that government and/or business will pull us back from the brink – that there will be a Great Awakening.

He talks about what will happen to various industries, especially coal and oil.

He says it is a near certainty that in a new economy dedicated to limiting global temperature rise to 1 (or even 2) degrees C above pre-industrial age level, coal and oil will be out. According to the German government funded Potsdam Institute for Climate Impact Research in 2024, we will have dumped all the CO_2 in the air we can. Any more CO_2 will increase the risk of going over the edge of climate disaster to greater than 20% - one in five.

So, assuming governments and business act, such a catastrophe will not be allowed to happen, which means that fossil fuels can no longer be burned after a few more years. (He discusses carbon-capture and storage - CCS - as a way to continue to use coal. Although he sees no reason not to investigate this technology he doesn't think it will prove economically competitive compared to renewable energy sources.)

It is less certain what fuel or technology will replace fossil fuel. Gilding doesn't think it will be nuclear because of problems with waste, terrorism and supply limits. He thinks it will be some combination of various renewable energy sources - hydro, solar, wind, geothermal. Right now, except for hydro, none of these technologies is much beyond the prototype stage. He believes that the economics of one or a combination of these approaches will be proven.

However, in the end, he points out, it is not about economics. Things will change. We will either go off an ecological/economic cliff - in which case, the change will happen to us. Or we will become the active agents of change.

AFTER THE GREAT DISRUPTION - DEALING WITH GROWTH
The Great Disruption is just the beginning.

Even after we have...

- Awakened to the climate crisis,
- Gotten on a war footing,
- Fixed the climate crisis (because we are good at war),
- Suffered the disruptions caused by shifting to renewable energy sources (necessary to fix the climate crisis),
- Reaped the economic opportunities created by shifting to renewables,
- Celebrated the cool new technologies invented to implement the shift

...we will still have to deal with material growth.

The climate crisis is only a symptom of the underlying problem. We live on a finite planet with finite resources. We can have spiritual growth, intellectual growth, moral growth - but not unlimited material growth. Physics doesn't work that way.

Gilding acknowledges that...

- The transition from material growth will be very painful. Economic growth is the foundation of our market-based system. It is the standard by which we judge our economy and our governments.
- For most of the 20th century our market-driven, growth based system worked. A lot of poor people moved into the middle class. A lot of middle class people became rich. (And a lot of poor people stayed poor or got poorer.)
- Even today the market based system still seems to be working. But this is a mirage. As our visible wealth increases, our hidden social and environmental costs increase even faster. At the macro level we are losing rather than gaining

wealth. Like the overextended homeowners of the 2008 financial crisis we are living in the midst of a giant Ponzi or check kiting scheme that is about to go bust.

However, even if we could have unlimited growth (which we can't) it would not make people happy. Over a certain level of income (at the time of this writing about $15K/year for individuals, $60K/year for families) studies indicate that additional income does not correlate with increased happiness. There is passing satisfaction at the moment money is spent but no lasting pleasure. The new thing becomes the old thing. People do get satisfaction from making more money than their peers but it is not the absolute income that matters only the difference compared to somebody else.

And although some rich people claim that raising the wealth level for rich people also raises the level for poor people - at least allowing them to move into the 15K "happy bracket", research shows that increasing inequity within a society degrades the quality of life for all citizens.

GILDING CLOSES WITH A STORY AND A POEM
Grabbing for failure.

Gilding imagines he is sitting in a cafe in Amsterdam in 1938 with his friend Pieter. Across the canal is the Frank house where Anne is now nine years old. Pieter, speaking with prescience , says that he believes in a few years Germany will violate Holland's neutrality, occupy most of Europe, kill maybe six million Jews (including the little Frank girl across the way) in a war that will eventually claim fifty million civilian and military lives. He further speculates that the US will not get in the conflict until the last moment - when the issue is very much in doubt.

Although Pieter feels the danger is obvious he doesn't see the possibility of any action taken anytime soon - even in Europe which will be affected first. The political leadership is not there; neither are the people.

As the now depressed 1938 Gilding cycles home, he wonders what he should do. That night he discusses the situation with his wife. They agree; it can't be that bad. Surely the leadership will do something. They decide to wait and see what happens.

Modern day Gilding's point is that the leadership is us. Business leaders (many of whom Gilding regards as smart, moral people) are constrained. Political leaders (many of whom who are smart and moral; many of whom who are idiots and immoral) are also constrained.

It's easy to give up. Gilding quotes this from the poem Common Sense by Paul Williams...

On the edge of the dream
we face our deepest doubts.
Now that it all is almost real
a terrible fear of success takes hold
and we grab desperately, uncontrollably, for failure.
One last chance to get off easy.
Who among us really wants to save the world,
to be born again into two thousand more years
of struggle? How much sweeter to be the doomed generation,
floating gently on the errors and villainy of others,
towards some glorious apocalypse now...
Hallelujah! It's not my fault – Bring on the end times!

Gilding's final point is this.

We need to get past stuff. We need to live happy, meaningful lives, not lives dominated by empty consumption.

Further, we need to talk and act - to get involved, even get mad - but as Gilding notes (maybe hard for some of us) not to get crazy. We need to follow the examples of Nelson Mandela and Martin Luther

Great Disruption   collapse of consumption   Gilding

THE UPSIDE OF DOWN: CATASROPHE, CREATIVITY AND THE RENEWAL OF CIVILIZATION

Shearwater Books 2006

ISBN 978-1-59726-065-7

Thomas F Homer-Dixon,

The curve is complexity.

The demand focused on by Homer-Dixon is the pursuit of energy needed to fuel growth. Supply is the source of available energy. The ever increasing search for new sources of energy results in an ever increasing complexity of systems used in the search – which according to Homer-Dixon makes the systems more fragile and vulnerable to a fall. Eventually, when demand exceeds supply and systems become unmanageable, the complexity curve starts to wobble, and a pulse happens (what Gilding calls a Disruption). If we pay attention to the pulse and control demand, the ultimate collapse can be avoided.

SUMMARY

The theme of the first book distilled in Crossing Infinity – The Great Disruption - is that unlimited growth is happening on a finite planet. The growth can't go on forever. Our resources are being up.

A principal theme of The Upside of Down is that unlimited growth results in the loss of cheap, readily available energy. Energy is a critical resource; all other resources depend on it. In order to maintain energy, ever more complex systems must be devised, to...

- Find and extract new sources of energy
- Make dirty – but available energy - environmentally safe
- Develop more efficient means for using any kind of energy

Besides being more expensive these complex techniques become more connected in a flat earth. An event in one place triggers an event in another place which, feeding back through the web of connected nodes, accelerates the pace of change in a nonlinear fashion. Systems become more fragile to a disruption, a pulse, a fall.

SOCIETAL PLATES

Big chunks of society are like tectonic plates. Rubbing together results in stresses which must be relived.

Human societies are threatened by interrelated stresses:

1. Population – The growth rate is different in rich and poor societies. It has peaked in some rich societies. The poor flood into rapidly growing megacities (e.g., Dhaka in Bangladesh). Conflicts between and within various regions (crowded and open) creates stress. Population pressures affect energy use, the environment, the climate, and economics.
2. Energy – The high quality energy (primarily oil) that fuels growth has peaked – we are now scrambling for less available more expensive oil. The scramble to get energy and maintain economic growth creates stress. Energy extraction and consumption affects the environment and the climate.
3. Environment – The natural environment is being destroyed. Resources are being used up. The struggle between those who exploit and those who conserve resources creates stress. The state of the environment affects the kind and quantity of energy available, the economy and the climate.
4. Climate – The atmosphere is changing; the planet is warming. Climate change creates stresses in all other societal plates.
5. Economic –The gap between rich and poor is widening. Societies are becoming unstable, prone to revolution and terrorism. Conflicts between rich and poor create stresses. Activities in all other societal plates affect the economic plate.

These conditions are like tectonic plates bumping into each other sometimes sliding, sometimes sticking building up stresses which ultimately must be relieved. These are societal plates.

Relieving stress results in major social transformation "pulses", comparable to the transformation from hunter-gatherer to agricultural society, the industrial revolution, and the communications revolution.

After the pulse there will be an opportunity for renewal – or utter destruction.

PLATE 1 - POPULATIONS
People are the driving force that makes everything else happen.

Rich countries have solved some visible signs of pollution (we dispose of trash, replant vegetation on old strip mines, etc.). Our farms are more efficient. So are our factories. However, our demand for energy and resources drives the rest of the world. We push environmentally damaging practices to poorer countries.

Poor countries don't have the money and education to pursue environmentally sound approaches. Masses of people living on the edge don't have the means to move away from the edge. That's where they stay – closer and closer to the edge.

There is a difference in birth rates between rich and poor countries. Birth rates in rich regions (Western, Northern Europe, Japan, white US) have stabilized or declined. Birth rates in poor regions are exploding and will continue to do so for decades (unless The Fall happens sooner rather than later).

Both rich and poor regions grow in order to maintain thermodynamic imbalance (the constant inflow of high EROI energy). However, growth is different in rich and poor regions.

In rich regions with higher levels of technology growth seems more benign. Opportunities are created. Interesting complexity happens when people of different backgrounds come together in rich cities. Culture flourishes.

In poor regions, poverty is perpetuated as more and more people must be supported by inadequate, dysfunctional social and technical systems. People tend to congregate in mega cities. Crime flourishes. The environment is destroyed. Somalia and Haiti happen.

Era of Anthropocene (Effects of People)

We live in a new geologic era – where the effects of human activity dominate and threaten all life, including the humans responsible for it. Human activity underlies everything. Homer-Dixon calls this era the anthropocene.

A lot has happened in the anthropocene.

Oxidation of organic matter, mostly fossil material that was formerly sequestered beneath the ground, has increased one hundred fold in the past several centuries. This has resulted in a corresponding increase carbon dioxide and a corresponding increase in global warming.

The same processes have increased the amount of sulfur and nitrogen released into environment. Both these chemicals affect the planet's chemical makeup. Nitrogen effects plant growth. Algae blooms sink to bottom of lakes, oceans. When plants decay they suck oxygen from water, creating dead zones where other life cannot survive.

We've changed the landscape - destroyed forests.

We have killed other species - animals and plants, on both land and in oceans. Having depleted top-of-the-food chain predator fish we eat more of the lesser fish – that are bonier and oilier. Homer-Dixon says that almost all fish have been taken from Mexico's Gulf of California.

Hollow Societies

Even societies that seem to be doing well are hollow. They are using resources from the future and from other places. They are empty shells.

Denial

Not all believe the anthropocene is real. There are three aspects of denial:

- Existential Denial – We say the problem does not exist; it has been manufactured by eco extremists and anti-capitalists.
- Consequential Denial – We say the problem has been exaggerated and the problems that do exist (e.g., loss of artic ice) do not really affect us. It's too bad for the polar bears.
- Fatalistic Denial – We say there is a serious problem but we can't do anything about it so we might as well live as we have been living all along.

PLATE 2 - ENERGY

Energy makes stuff go. It makes modern complex societies possible. It makes us – as we see ourselves now - possible. The loss of it can destroy us.

EROI and High Quality Energy

High quality energy is energy whose return on investment - it's EROI - is much greater than one. In other words, the energy brings in a lot more than it costs to get.

Generally, high EROI energy comes from fossil fuels - oil, coal, gas. The supply of these fuels is by nature, finite.

Societies require high quality energy to grow and prosper, to become complex. Conversely, societies need to grow and proposer - to become complex - in order to take advantage of ever diminishing sources of high EROI energy. A society has to be smart, wealthy and technically savvy, in other words, complex, to get high quality EROI fuel (especially oil) from ever more inaccessible locations.

Peak Oil

Oil has been the world's preeminent energy source since the early 20th century (the first commercial oil well was drilled in Pennsylvania in 1859). For most of the 20th century the EROI (energy return on investment) of oil has been high. However as the supplies of readily available oil are found and as production peaks the cost inevitably goes up. In U.S from 1930 to 2000 EROI dropped from 100 to 1 to about 17 to 1.

U.S. oil discovery peaked in 1930. Following predictions (called Hubbert's curve) U.S. production peaked in 1970. This happens when about half of available oil is extracted.

World oil discovery peaked in the early 1960's. According to Hubbert's curve and according to data cited by Homer-Dixon, world production peaked around 2000.

Most of readily available oil is found in a limited number of giant oil fields. Many of these are in the middle east. No new giant oil fields have been discovered in years. The only new oil found these days is in difficult places or is not high quality. The EROI is low.

The notion of limited oil reserves and inevitably declining production is called Peak Oil. Some experts claim that the concept of Peak Oil is bogus – that there will always be more oil. Homer-Dixon says even if this were true, the EROI on newly discovered oil inevitably will go down the as the cost to extract and process goes up.

When Peak Oil takes effect we'll have to switch to other sources of energy – natural gas, coal, wind, solar, nuclear, etc. We will also have to become more efficient. However each of the alternate energy sources has problems, some to do with EROI, some to do with pollution (the control of which reduces EROI). With possible exceptions of ethanol and coal, Homer-Dixon doesn't think there will be enough energy available to offset the loss oil. There will be a Fall.

Thermodynamic Entropy

As thermodynamic system, societies, using high EROI energy, become growing islands of order in ever increasing seas of entropy (disorder). Societies create these disordered regions as they grow. As societies get bigger and more complex, they require more and more high EROI energy.

Growing societies (all complex adaptive systems, as explained later) are in a state of thermodynamic imbalance. They are not in a state of equilibrium with their surroundings. At some point the EROI starts to fall, eventually, inevitably approaching an EROI of one (actually about four). At that point, the society succumbs to the thermodynamic imbalance – to the pressure of entropy – it collapses.   Entropy eases over the debris like still water. (In asymptotic terms, the curve collapses because the complexity requited to maintain such a system falls.)

PLATE 3 - ENVIRONMENT

The environment is being destroyed by the growth of human consumption.

The growth imperative fuels environmental problems because growing societies, even in "post-modern" countries tend to increase overall consumption and pollution, even though some local pollution diminishes (often exported to third world countries).

Alternate view: Conservative economists say that as industry changes, we become ever more efficient, ever smarter. We pollute less, use less. We rely less on the physical world, more and more on a world of our own making. Homer-Dixon counters that growth has cancelled out these gains. It will always cancel them out. Our energy footprint gets bigger. Also, much of our polluting industry has been moved to other regions of the Flat Earth, which can (because the earth is flat) still get us.

PLATE 4 - CLIMATE

Homer-Dixon regards global warming and resulting climate change as one of the societal plates underlying civilization. It can bump into other plates (the environment, economy, energy, and population) and get stuck – then break loose with disastrous results to all the other plates.

How Planetary Temperature Maintained

Planetary temperature is an underlying factor of climate.

Planetary temperature is a balance between radiation coming from the sun and radiation emitted back into space. The balance changes, the planet gets hotter or cooler, depending what's going on in atmosphere.

For example, ash spewed into the atmosphere from a volcano can restrict radiation received from the sun and make the planet cooler. Changes in the sun spot cycles can do the same thing.

If the atmosphere absorbs more radiation in heat-absorbing frequencies, the planet gets hotter. Gases that absorb heat radiation include carbon dioxide, nitrous oxide, methane - all of which are produced in quantity by human activity.

Controversy

There is some legitimate scientific controversy regarding global warming. However much controversy is the result of bogus arguments presented by non-scientists or by a few fringe scientists. Media sources, not understanding science, attempt to be "fair" by presenting both sides as if they are the same.

Levels of Certainty

This is what we (science) knows with confidence:

- The levels of greenhouse gases have greatly increases since the start of the industrial revolution 150 years ago. Carbon dioxide has risen by a third, methane by 100%.

- <u>The average temperature</u> of the earth's surface has risen 0.8 C since the last half of nineteenth century. The rate of warning since 1976 has tripled the increase in the previous century.

We are somewhat less certain of the following:

- <u>We are not absolutely sure if global warming is responsible</u> for various weather anomalies (increased numbers/severity of hurricanes, droughts, etc.). Systems are too complicated to be explained by single causes.
- <u>We are not absolutely sure if increases in greenhouse gases</u> have caused observed increase in planet temperature. However, models – which are disputed by skeptics – show a correlation. These same models hold up when applied to historical data.

Factors Affecting Climate's Future

These are the factors:

- <u>Momentum</u> – Stopping or even slowing the release of greenhouse gases will be difficult. Momentum caused by increases in population, fossil fuel consumption, loss of environmental diversity, etc. will have to be overcome. This will be especially true in developing countries like China and India.
- <u>Feedback Loops</u> – Some effects of climate change could result in even more climate change. Consider these examples:
    - The loss of artic ice could expose darker seawater which absorbs more heat which accelerates the melting of ice which causes the temperature to rise.
    - As permafrost in Siberia melts it exposes peat. When peat decays (upon exposure to the atmosphere), it releases methane which is a greenhouse gas which accelerates global warming.
- <u>Equilibrium Shifts </u> - Geo-systems can shift from one equilibrium to another. Systems can have multiple equilibrium points acting like adjacent basins. Systems can slip rapidly between equilibriums. The system never stays between basins – it is always in one basin or the other. No equilibrium is preferred over another equilibrium.

It is the stuff of science fiction.

Consider the Gulf Stream section of the so-called "great conveyor". The Gulf Stream is an equilibrium point – one of the basins. Tropical sea water is heated by the sun. It becomes more saline. This warmer , more saline water is driven by surface winds North where heat is released. Heating keeps Northern Europe and the British isles habitable.

Once cooled, the water becomes heavier and sinks to the ocean depths. It flows South. The flow of the great conveyor keeps heat in equilibrium around the planet.

However, global warming seems to be decreasing the salinity of sea water. This could affect the operation of the conveyor. It could shift – with little warning - to a new equilibrium. The conveyor could stop working all together. Thus while the planet as a whole gets warmer the North Atlantic regions could get much cooler with dramatic effects.

PLATE 5 – ECONOMIC

As always, money means a lot.

Citing a debate in 1997 between currency speculator and rich liberal George Soros and economic commentator (also liberal) Paul Krugman, Homer-Dixon questions whether capitalism (e.g. the competitive marketplace) can result in economic equilibrium – e.g., in stability. Soros and Homer-Dixon argue that competitive economic systems are inherently unstable, influenced by the subjective opinions of traders and speculators. Such systems, the two men claim, tend toward extremes – achieving equilibrium only in intermittent periods between market rises and falls.

Note on Soros – He uses the term "reflexivity" to describe the notion that the biases, prejudices, and ideas of investors tend to alter perception of economy. Trends and market movements become exaggerated. Investor biases become self-fulfilling prophecies. Equilibrium never happens.

The market, Homer-Dixon says, is nonlinear. In the manner described by Soros, feedback loops happen. Actions begat reactions which feed back into actions which begat more reactions and so on. Such actions are exacerbated by the connectivity of modern financial systems. An event on one side of the globe has an immediate effect on the other side of the globe which feeds back into the original action and so on.

It is the flat earth at work.

Modern capitalist economies depend on ever increasing consumption. People are exhorted to buy more and more stuff – to fill the empty holes in their lives – and (implicitly stated) to support businesses that were created in anticipation of ever increasing consumption.

We are caught on a "hedonic treadmill" – forever seeking more and more stuff. This is true even though studies have shown that past a certain point added wealth does not make us happier. We do not seem able imagine anything else. Although, Homer-Dixon points out, that might not always be true.

The Rich and Poor Gap

The gap between rich and poor is also exacerbated by growth. Although the rate of income growth might be greater in some poorer countries, their people started so far behind that the rich are still getting richer. It is another nonlinear effect.

Homer-Dixon points out that even though the gap between rich and poor gets bigger and bigger the standard of a good life as measured by consumerism is becoming more universal.

Poor people know they are poor, unequal.

(See the asymptotic premise of Coming Apart.)

Pursuing Growth

Modern capitalist societies depend on growth. Responding to opportunities and the desire for wealth, entrepreneurs create new businesses.

Growth is necessary to support the results of previous growth. Growth and demand hardly ever match. Growth gets ahead of and then falls behind demand. Pressures to grow exacerbates boom and bust swings.

Pursuing Productivity

Growth is also a by-product of productivity.

One way to grow is to become more efficient – to do the same with less. Responding to competition (from home and abroad) businesses constantly search for productivity and efficiency.

Displacing Workers

Pursuing productivity affects people. Becoming more efficient means displacing workers.

Consequently, growth must happen to give these people jobs so they can buy more stuff and maintain the economy – and the growth that has already happened.

However, chasing productivity is like chasing your own tail. It stops when something breaks.

This is the fear.

At some point, there simply might be no place for displaced workers. The lower skill jobs which once absorbed these people go away – through automation or offshoring. A new class of permanently underemployed people, resentful and dangerous might is created.

The house of cards falls in. Revolution happens.

(Again, see the asymptotic premise of Coming Apart.)

Alternate view: Many economists and experts say that through training and development, people are constantly moving up into higher skilled, better paying jobs. Evidence and current tendencies to let people fend for themselves refute this view.

COMPLEXITY IN RESPOSE TO STRESS

Stress results in the movement of societal plates which occurs as growth happens. Complexity results from the development of systems designed to alleviate stress. Dealing with complexity creates even more stress which results in even more complexity which accelerates in a nonlinear manner. Connectivity ensures that activity in one plate is felt by other plates.

Complexity cannot be sustained. A fall must happen. But there can be renewal after death.

The next sections describe complexity, connectivity, and nonlinearity in various ways. .

Complex Adaptive Systems

Human societies are complex adaptive systems. So are animal societies, the stock market, the biosphere, all businesses and human minds individually and collectively. The list goes on. Characteristics of complex adaptive systems include connectivity, nonlinear reactions (small causes can have large results), energy usage and growth.

The last two characteristics are especially important. Complex adaptive systems are thermodynamic. They suck energy out of environments, consuming the readily available energy first, then as that supply is exhausted, the more costly energy (in terms of money and effort to obtain). Adaptive systems that are smart enough try various compensating tricks as energy becomes scarcer. For example, in burgeoning rat populations, the strong eat the weak - which might be analogous to what some rich humans do to some poor humans.

Homer-Dixon says human societies adapt to resource scarcity by becoming more and more complex, more connected and interdependent. We squeeze every last bit of efficiency out of our systems, until there is nothing left to squeeze. In the process we lose resiliency, become fragile, subject to disruption.

The final stage of the growth cycle is common to all adaptive systems. The system becomes increasingly vulnerable to black swan events. Inevitably something happens. The whole thing falls apart. The Lemmings go off the cliff. The rat population collapses. The locusts, having eaten everything, starve. The forest burns. Rome falls.

There is a Fall.

We fall into the comfort of simplicity. We no longer bear the burden of maintaining our complex systems. The asymptotic curve topples. .

Joseph Tainter and Complexity

By adapting current systems to new data (e.g., adding epicycles within epicycles) we increase the complexity of the current systems.

Joseph Tainter is one of the founders of the view of societies as complex systems. He described his theories in the 1988 book *The Collapse of Complex Societies*. His basic idea is that societies become complex as they adapt and change to solve problems.

For a time this works, depending on how adaptable and creative the society is. The society grows, prospers. New energy is added to the system.

However, developing and maintaining complex solutions requires additional layers of description and control. These layers cost – which means that more and more energy is required. Solutions become increasingly convoluted, consuming ever more energy. (Considering a household as a thermodynamic system, over the years more and more energy is required to implement the "solutions" needed to maintain the household. Or consider a large software project. Over time the basic programming team is expanded to include project managers, business analysts, technical writers, quality assurance testers, marketing people, etc.)

Up to a point complex solutions might increase the amount of energy available. Past that point the solutions consume more energy than they produce. The society goes into the hole and unless it stops whatever it is doing must ultimately fall.

According to Tainter, societies stick with complexity because short-term solutions tend to acquire momentum. The costs add up.

He also notes that there are more "concatenating" problems. These are issues (and solutions) that tend to interact in unexpected ways.

Crawford Holling and Panarchy

Holling is Canadian ecologist who worked in forest management. He founded a group called the Resilience Alliance. It promotes "panarchy" (from Pan, the Greek god of nature) – as opposed to anarchy.

Panarchy theory says natural systems (and societies) operate in adaptive cycles. Such systems grow, become more complex as available niches (ecological and social) are occupied. The system becomes thermodynamically unbalanced. It takes more and more energy to maintain in a highly adapted condition.

As a system grows, its parts become connected (in a forest, via bugs, bacteria - in human society via electronic media). These connected parts regulate the system, control its operation.

Connectivity and regulation affect the system's resilience. Early in its growth phase, the system has great capacity for change and novelty. But as its parts connect and regulate one another, the system tends to become more rigid. Eventually something happens – a black swan for instance – and the system, unable to respond to the situation, collapses.

This is the adaptive cycle.

Often the collapse of a cycle though painful is beneficial – giving the system an opportunity to grow back stronger, different. Think Pearl Harbor.

Adaptive cycles occur at all levels in a system, each level having its own cycle. Some cycles might take seconds or minutes to go from start to collapse (the development of bacteria in a forest) some might take weeks or months (cycles of activity on world wide web) some cycles might take decades or centuries (the lifetime of cultures). In many cases, cycles are embedded in cycles. That is what Holling calls panarchy.

Generally collapses at one level do not coincide with collapses at other level. This promotes stability, what Holling calls resilience.

When multiple parts of a system fail at one time it is called a deep collapse.

The world is now in danger of a deep collapse because globalization and technology have connected all its parts. If one part goes, they might all go. We have maintained our growth phase in the adaptive cycle, become more complex, more rigid, more dynamically imbalanced.

We are not resilient.

Recognizing Complexity and Panarchy When We See It

We often do not recognize our complex adaptive systems for what they are – do not see panarchy at work.

We stick with existing theories as long as we can. Rather than adopt new theories, we adjust old theories to accommodate new data. We layer on complexity.

Homer-Dixon notes as an example the shift from the geocentric view (per Ptolemy) to the heliocentric view (per Copernicus, Kepler, Galileo). To explain the data that Galileo discovered with his new telescope, proponents of the old geocentric theory devised elaborate models (called armillary spheres) which showed heavenly bodies rotating around the earth in cycles within cycles.

We resist changing basic ideas and assumptions for various reasons...

- We don't recognize some problems because of the time scales involved. Called "slow creep" issues, these problems develop so slowly we can't see them. Some of us just can't imagine "deep time" – or anything not directly in front of us.
- We prefer to stick with what we know.
- We have a vested interest in the status quo. We make money from it or are defined by it. It is part of our script.
- We are scared.

Shifting to a new view can be disorienting, challenging to common sense. It seems "obvious" that the sun moves across the sky.

Accepting that it is actually the earth rotating requires a shift in point of view. It requires imagination

CONNECTIVITY

Connecting systems to one another allows things (material and immaterial) in one place to spread to another place. Good things and bad things can spread – inventions, ideas, wars, diseases, ecological disasters, and so on.

Random and Scale Free Connected Networks

Connectivity happens in natural and man-made networks, which are random or scale free. (Nassim Taleb also babbles about this in the Black Swan, which is distilled later in Crossing Infinity.)

Random networks are like interstate highway system; scale-free networks are like air-traffic system.

In random networks, nodes are more-or-less equally linked (graphed by a bell curve).

In scale–free networks (pictured by Pareto curves with big heads and little tails – like s bell curve shifted to one side), a limited number of nodes have a disproportionate share of links. These nodes are called hubs. The world wide web, electrical grids, food distribution networks, etc. are scale-free.

Up to a point scale-free networks handle disruption better than random networks. However, take out a critical hub and a scale-free network falls apart. Consequently scale-free networks are both robust and fragile.

We live in a complex world of interconnected scale-free networks. At some point the complexity and the sheer volume of traffic could bring these networks down.

Aside – The terms random, scale-free, scalable, and non-scalable are generally applied to networks – living and non-living. When used metaphorically they get tricky. The term scale-free was invented by Albert-Laszo Barabazsi in 1998 subsequent to his study of connectivity in the world wide web. He discovered that components of the web are not randomly connected but are scale-free networks. Both random and scale-free networks are scalable – they can grow and get bigger.

Plates and Quakes as Instruments of Connectivity

Tectonic plates are connected by proximity and friction. Sliding tectonic plates exert force against each other and get stuck. When the plates finally move there is a sudden release of energy. An earthquake happens.

Social quakes work in somewhat the same way. The societal plates that underlie civilization – energy, population, environment, climate, economic – rub against each another. They too are connected by proximity and friction. Changes in one plate create pressures and stresses in other plates. For example Peak Oil increases the cost of gasoline which exacerbates the growing gap between rich and poor. Population growth – especially in poorer regions increases pressures on the environment. Peak oil also affects climate, as we are forced to use more polluting fuel. Stresses build up until something snaps: a revolution, the collapse of an environment, a recession.

High levels of connectivity exacerbate breakdowns. Changes in one place are rapidly experienced somewhere else allowing the nonlinear processes described next to build up. Flexible, adaptable societies are better able to handle stresses created when societal plates rub against another and get stuck. The most adaptable societies are typically western-style capitalist democracies.

NONLINEAR EFFECTS AND NEGATIVE SYNERGY

Nonlinear events spread rapidly across connected networks. This affects that which affects this in a feedback loop. Recall that an aspect of asymptotic curves is nonlinearity

Nonlinearity is the feedback process by which societal quakes are spread.

Negative synergy is the result.

Synergy happens when multiple events combine to create non-proportional output. Positive synergy is when the results are greater than the input – when 2+2 = 5. Negative synergy is when the results are less than the input – when 2+2 = 3.

The rubbing of societal plates to create societal quakes results in negative synergy. Bad things are propagated in a nonlinear manner. All of the societal plates listed above tend to have a negative synergy. Conditions in one plate jump boundaries to another plate and

ROME AS AN EXAMPLE

What happened before.

Admitting that there are already 200 explanations for the fall of Rome Homer-Dixon offers his explanations - setting up a comparison between that fall and the one that threatens us.

Regarding the fall of Rome, he notes...

Rome expanded by mining the energy of conquered people –as stolen gold and silver, food imported and taxes levied. Elaborate networks of Roman roads facilitated various kinds of traffic between the center and provinces:

- <u>Movement of military forces</u> from Rome to where they were needed.
- <u>Transport of food</u> (energy) from the provinces to the center.
- <u>Flow of information</u>.

Rules, regulations – systems were developed to control the empire. Complexity increased.

During the first centuries of the Roman Empire the rate of return on the cost of conquest was high (the EROI - energy return on investment - described in "Plate 2 – Energy" was much greater than one). The empire was profitable. Because the empire was made of so many parts, problems in one region could be compensated for in other regions. In terms of adaptive systems, the empire was resilient.

However, eventually the empire spread too far. Its systems (physical and bureaucratic) were stretched too thin. The EROI started to slip.

For decades, new rules and controls, and additional taxes were imposed to offset the energy deficit and to the keep the system going. However the cost of energy still went up as the supply of energy (food, resources) fell. The EROI dropped below the break-even point. Becoming ever more complex and pervasive, controlling systems became ever more rigid and inflexible. Again, in terms of adaptive systems, the empire was no longer resilient.

The empire was in a state of thermodynamic crisis. It was an accident (a conquest) waiting to happen, which history provided as the barbarians from the North.

There was a collapse.

NEXT (THERE WILL BE A PULSE)
What happens next.

We (the modern world) are at a point similar to the one faced by Rome – late in the growth phase of the adaptive cycle. The solutions we pursue indicate where we are.

All organizations (business and governmental)...

- <u>Become more and more complex</u> as they try to find better and more efficient ways to handle scarcity and competition (many companies go after the same base).
- <u>Increase productivity</u> (trying to get the most out of each worker through automation (technology) and by offshoring labor to cheaper countries.
- <u>Exploit energy sources</u> that previously would have not been considered economically viable – now such sources are increasingly all we have left.

Major Pulse Awaits

Enmeshed in complex solutions to diminishing EROI, we are losing resilience. We are entering a period of extreme vulnerability when a major "pulse" could happen. Previous pulses include...

- The transition from hunter-gatherer to agriculture society.
- The industrial revolution.
- The global communications revolution.

Further indicators include the following...

- <u>Foreshocks</u> becoming increasingly severe, more frequent. Examples include the economic downturn of 2008 and the "Arab Spring" of 2011.
- <u>The gap</u> between the rich and the poor increasing. The middle class gets smaller (see Murray's *Coming Apart*). Marginalized young men – seeing the gap between the rich and poor and the powerful and powerless, become more active. They might gain access to nukes. If Saudi Arabia goes we go.
- <u>Environmental disasters</u> , aggravated by global warming, happening.

Three Places Where Pulse Could Start

Homer-Dixon's best candidates are...

- Saudi Arabia – It is vulnerable to political upheaval, from within or from the region. When the Peak Oil happens in the next decade or so, production will drop. This will have major effect in rest of world.
- Pakistan – They've got nukes; they are unstable for various reasons, but especially due to the gap between rich and poor.
- China – The worry isn't that China will get too strong, but that its weaknesses will bring it down and because it is so huge it will bring down the rest of world. The four challenges for China are:
  1. Huge population – Many of China's people are still poor. As people leave the countryside, they concentrate in big cities. People aspire to First World lives while stuck in Third World conditions. Great inequality exists. (See the Happiness curve in Coming Apart.)
  2. Inefficient polluting industries – China's industries are vulnerable to disruptions.
  3. Dubious government – Governmental entities left over from old communist regime are prone to failure.
  4. Degraded damaged resources, natural environment - China is increasingly becoming an importer of resources. It is vulnerable to environmental disasters, famine. China must grow to accommodate migration of people from countryside to cities and factories. This makes the country vulnerable to foreign markets. China is especially tied to US. If US has a recession, China suffers. (Both countries must grow.) China lends money to US banks. If they fail, China feels it.

Europe also a Dangerous Place

Native populations of European countries decline while the immigrant populations increase. The immigrants are typically poorer, which creates tensions between rich and poor. (Again see *Coming Apart*.)

The energy resources of European counties are declining which makes the region more vulnerable to the vicissitudes of Middle East suppliers.

Also there is the danger that the North Atlantic conveyor of warm water from tropics (the Gulf Stream) could be disrupted, making Europe a lot colder.

Moments of Contingency

These moments happen late in a growth cycle, just after a social quake. Everything becomes contingent. Nothing is definite. The center does not hold. Small actions in one place can yield large results somewhere else.

Homer-Dixon thinks that the course of events depends on how leaders frame the issues, rendering the complexity into simple terms, exploiting the situation for evil or good ends.

RESPONSE

We've been lucky – avoiding nuclear war, escaping plagues, not triggering the worst environmental disasters. But as EROI drops, plates will get stuck, pressures will build up. A Fall will happen. Luck cannot hold.

HOMER-DIXON'S CATAGENESIS

Our immediate responses will not work. We need to embrace change.

Our response to an impending Fall (to sub-Falls, not the Great Fall to come) has been twofold.

- We deny the problem exists.
- We try to manage the problem by maintaining the status quo – the current system.

But management only makes our systems increasingly complex. It diminishes our resilience. We become more vulnerable.

Homer-Dixon says we need to embrace change – even a Fall. We should try to come out better on the other side.

He says we need to develop prospective minds.

He calls the process of renewal Catagenesis.

Proposed Actions

Homer-Dixon proposes four actions – taken before and after social quakes:

Reduce Tectonic Stress.

Try to solve problems before collapses happen. Aggressively move on multiple fronts, crossing boundaries, not getting caught up in single silos. But this approach is tough, especially for slow creep problems. People just don't see them

Cultivate Prospective Minds.

Nonlinear processes and threshold events make the future impossible to predict (See Taleb's Black Swan). Science can help distinguish between plausible and non-plausible scenarios. But in the end we need to have prospective minds that are capable of handling change, surprise, and profound uncertainty.

Boost Resilience of Critical Systems.

We need to make critical systems resistant to total collapse. Homer-Dixon says a principal way to do this is to sacrifice some efficiency and productivity for reliability and resilience. Don't make everything contingent on everything else. That way one failure does not ripple through the entire system. Distribute command and control through the system. Rely on a collection of emergent control subsystems instead of one central control. Employ semi-independent nodes. Homer-Dixon says resilience is an emergent property of an entire system. In our individual lives we should try to become resilient and self-sufficient. We should pursue lives that are less encumbered.

Turn Breakdown to Our Advantage.

This is catagagenesis – the rebuilding and renewal that takes place after a Fall.

We need to allow for breakdown, to expect it and plan for it. We need to be orderly but not too orderly, careful but not too careful. When the inevitable breakdown happens it should be contained and constrained – so that all the layers of society do not pancake in on one another. We need to decouple parts of society as much as possible to minimize nonlinear effects – so that one part doesn't bring down the whole thing.

(Homer-Dixon suggests that rich societies wean themselves away from consumerism. He also warns that extremists will always have the upper hand over moderates – the extremists will be better organized, more committed.)

Homer-Dixon notes three characteristics cited by scientists of highly adaptive systems:

1. Such systems are comprised of many <u>diverse parts</u>.

2. Decision making and problem solving are <u>distributed across systems</u>. Information is shared.

3. Such systems are <u>loose enough</u> to come up with unexpected solutions but <u>tight enough</u> to share the solutions.

FINDING VALUES
To survive the coming Fall we need to examine our values.

We need to move beyond strictly utilitarian values which only express our likes, dislikes. This gives rise to our consumer oriented culture. Consumed by consumerism, we become less resilient, more vulnerable to unexpected nonlinear events.

We also need to move beyond the unthinking, politically oriented notions of fairness and right and wrong which often passes for spiritual belief. This too leaves us rigid, vulnerable to nonlinear events.

We need to move into the realm of spiritual and existential values that are "compatible with the exigencies of the natural world".

According to Homer-Dixon, such values recognize that...

1. Energy and the laws of thermodynamics play a key role in our survival.
2. Certain kinds of connectivity are dangerous.
3. Many natural systems (including all adaptive systems) behave in a nonlinear manner.

In the words of Nassim Taleb we live in Extremistan.

We need to move from a growth imperative to a resilience imperative.

COMING APART - THE STATE OF WHITE AMERICA,1960 - 2010

Random House (Crown Publishing), 2012

ISBN 978-0-307-45342-223-6

Charles Murray

Murray's notion of a collapse is when society comes apart as a result of increasingly large lower and upper classes pulling population from a shrinking middle class. Murray says that this movement is caused by changing class structure which in turn is caused by a loss of responsibility, freedom, and a diminished adherence to what he calls the founding father virtues. These changes are concentrated in the lower class but can spill over into other classes. That will result in a failure of leadership and a loss of values across all classes.

Of underlying importance to Murray is how these factors affect the happiness of people that comprise a class.

One of the major factors contributing to happiness is jobs.

Consider the lower class. There is an increasing demand for the ever-diminishing manual jobs that can be performed by the ever-increasing lower class. Supply is represented by availability of these jobs. People in the underclass are not happy.

Contrast the plight of the lower classes with the condition of the New upper classes. They tend toward "knowledge work" of which there is an increasing supply. This increase is due in part to the increase in complexity described in Upside of Down. These people are generally happy.

The supply (of jobs) and demand (by job seekers) can pictured in an asymptotic manner to create a happiness curve.

The asymptotic collapse happens when the discontent of the lower classes overwhelms contentment of the upper classes. This he says, will start in Europe when that model of the welfare state fails. In asymptotic terms, the European collapse will provide the shudder in the happiness needed to wake us up to the general collapse to come.

.

SUMMARY

This is what Murray talks about...

A major change has taken place in U.S. culture since 1960 – the fourth in a series of Great Awakening, the first starting in the mid 1770's. In the current awakening new social classes have come into existence and old classes have diminished. The distance between the classes is greater than ever before. The upper and lower classes have so little in common they can barely speak. The principles of the founding fathers (honesty, marriage, religiosity and industrious) are still more or less practiced by the upper classes but not so much by the lower classes.

The lower classes (white and black) are falling apart. If this spreads to the upper classes we are trouble. The issue is in doubt.

GREAT AWAKENINGS

Four movements transformed the country.

Citing Nobel Economist Robert Fogel in his book *The Fourth Great Awakening and the Future of Egalitarianism,* Murray notes four Great Awakenings:

The First Great Awakening

Took place between the mid 1720's and 1730. It set the stage for the American Revolution.

(Note that the effects of an awakening might not be felt until many years later. Movements started at the time of an awakening are still going on - spanning multiple awakenings.)

The Second Great Awakening

Started around 1800 and lasted until 1840. It led to the temperance movement, compulsory elementary education, abolition and the beginning of the women's suffrage movement.

The Third Great Awakening

Began around 1860s and went to early 1900s. "It laid the ethical basis for the reforms of the New Deal and later the Civil Rights movement." The welfare state is the product of this awakening. Liberal notions regarding the mutability of human nature, responsibility, free will and fairness were brought to the fore here.

The Fourth Great Awakening

Began around 1960 and continues until today. It was started by born again Christians - members of "enthusiastic" religions. Main line churches and Catholic institutions who share these evangelical beliefs are part of this awakening. Arrayed against these true believers are people who are the product of the third Great Awakening. The emphasis of the fourth Great Awakening is on the spiritual needs of the country - the third Great Awakening emphasizes the material needs.

CLASSES

How we are classified and divided into groups, some of whom cannot even begin to understand the others.

Since 1960 - approximately when the Fourth Great Awakening started, new classes have come into being and old classes have diminished. Of course, one should be aware that the distinction between classes is arbitrary and subjective, the boundary fluid, permeable. People move back and forth.

Nevertheless, one can generalize about these attributes as being distinguishing class factors...

- <u>Money</u> – how much you have.
- <u>Culture</u> (what you like and don't like, whether you drink Bud Light or Dos Equis or maybe a craft beer, whether you watch Jeopardy or Masterpiece Theater – or maybe no TV at all – and so on).
- <u>Power</u> – if you exercise control over other people.
- <u>How well you adhere to the virtues of the Founding Fathers</u> (unless perhaps you are an Elite).
- <u>And maybe most importantly, how well you are able to defer gratification to achieve greater rewards later on -</u> how well you are willing to endure pain and discomfort now. Another name for this quality is will power.

Narrow Elite

These people run and influence things on a national level. They are presidents of national and multi-national companies. They are influential politicians, film makers, writers, artists, scientists. In an ape society they would be the smartest, most powerful, most aggressive, most dangerous members of the tribe. They have some of the same attributes as sociopaths but hide it better.

Broad Elite

These are people who run and influence things on local and regional levels - state, city. Influential people in some local levels - depending on the level - might also be members of the Narrow Elite. For example the mayor of New York, the governor of California are probably members of the Narrow Elite.

Note: The next two upper classes are relatively new, establishing categories for the substantial numbers of knowledge workers - a group that was just coming into being in 1960 - at the beginning the Fourth Great Awakening. The distinction seems especially arbitrary but Murray seems to think it is worth noting.

New Upper Class

These people are nearer to the Broad Elite class than to the Middle class. They are professional people - knowledge and information workers. They are doctors, lawyers, successful managers, writers, professors, architects, high level system programmers and designers. All these people make money from their brains. Most have gone to college; most make (or at least spend) significant amounts of money. Most have a cultural patina (real or fabricated). They take cruises, drink latte, discuss wines, movies - maybe books. They go to museums and concerts - even if they don't always understand what is on display or being performed. Their lifestyles and interests are very different from the Middle and Working classes and especially the Lower class. Depending on how many generations of ancestors have been members of an Upper class (either of the Upper Classes), descendants might know little about the lower classes. The people live in different worlds. Upper class people tend to marry and associate with other upper class people which tends to increase the separation in a non-linear manner.

Upper Middle Class

Although much like the New Upper Class and aspiring to that level, these people are closer to the Middle class. The biggest difference between the two upper classes might be money and how far one is removed from the classes below. An Upper Middle class person might only be one generation removed from the Middle or Working Classes.

(The family of a New Upper Class person might have been at that level for multiple generations. Their parents might have been called professionals and probably had the money and the background to get their children in one of the elite schools.)

An Upper Middle Class person probably went to a second tier or lower school. The Upper Middle class might include application programmers, technicians with college degrees, family practitioners, pharmacists. newspaper and local TV reporters, technical writers, etc. Community college instructors, maybe high school teachers might be members of this class.

Aside - The Artist Class

Artists, writers, performers who have not yet "made it" are somewhat classless. They tend to associate with all classes but remain in contact with the classes from which they came because that is where their material comes from.

Middle Class

Murray doesn't say much about this class. However based on common experience one could imagine that these people are typically not knowledge workers. More likely they would be owners of small shops and business, manufacturing production mangers, nurses, police, soldiers, etc. These people might watch basic TV, eat regular food, drink beer (and if wine would not know that red and white require different kind of glasses) They would not be ashamed of going to cafeterias, fish camps, and bringing home Chinese takeout. Money might account for some of the difference between other classes. It is possible that in terms of culture people from the middle class might have more in common with the elites than the upper classes. One can easily imagine a small shop owner from the middle class making it big and moving into the broad elite – bringing along middle class values and culture. The term nouveau riche applies to such people. Being a knowledge worker in the Upper Class seems less likely to happen.

Working Class

These are the carpenters, electricians, plumbers, machinists, beauticians, etc. who work directly for the people in the Middle Class and possibly indirectly for the people in the Broad Elite class. Some working class people are artisans, some are artists, some perform repetitive manufacturing jobs. People employed in blue collar factory jobs are members of the working class.

Lower Class

Murray devotes much of his book to discussions of the Lower class and distinctions with the Upper classes. Both upper and lower classes have many members who were born in their class and stay there.

All the class differences between apply here. Lower class people have less money, a lower-class style culture, little power over others, reduced adherence to the Founding Father virtues and probably most important – a reduced willingness to defer gratification.

However the Lower Class and the classes above include migrants from adjoining classes. The Lower class mostly gets migrants from the Working and Middle classes who have migrated downward because of lost jobs (due to downsizing, automation, outsourcing, and in sourcing).

The Upper Classes (especially the Upper Middle class) get upwardly mobile migrants with cerebral intelligence from the Middle and Working classes.

Murray notes that in times past there were not lower classes as such, just poor working people who did not have jobs.

FOUNDING FATHER VIRTUES

The previous section describes what the classes are. This section describes how they came to be – especially how the lower class became lower.

Murray blames the rise of the lower class as a failure to adhere to the virtues of "founding fathers". He blames that failure on the emergence of the welfare state as envisioned in the Third Great Awakening - which continued and accelerated in the 1960's.

Citing statistics to support his points, Murray explains the virtues of the founding fathers and how the lower classes fail to live to live up to these virtues...

Marriage

Marriage is the glue that holds families together, that ensures a successful next generation.

Members of the lower class...

- Tend to marry less, have more children out of wedlock.
- Are unhappier in marriage.
- Have fewer children living with both biological parents.

Murray quotes George Gilden who says in his book Sexual Suicide "men are barbarians who must be civilized by women".

Industriousness

Doing work, accomplishing something makes people feel good about themselves.

(The opposite makes people feel bad about themselves. That leads to unhappiness which is the name given to the asymptotic curve that describes an underlying theme of this book – which Murray might not recognize..)

Workers reduce the cost to the rest of society (Murray's seeming response to the plight of workers with one or more jobs who still can't get by is that by the standards of times past they are getting by and anyway, people are not the same and unfortunately some will always do better than others.)

In the lower class...

- The number of people receiving disability payments has increased, although the actual rates of disability have gone down.
- Fewer people say their jobs are worth doing - that they feel a sense of accomplishment. They prefer shorter hours and security to chances for advancement.
- Lower education levels reduce the availability for work among the lower classes (not necessarily those working but those willing to work).
- Male unemployment is greater

Murray aside

Some people blame the quality of jobs, the lack of jobs - the labor market itself for their difficulties. He says that men today can afford the luxury of having these attitudes because they don't have to work to survive. They don't have to support children. They get welfare, or live off the food stamps of others - they are on the dole.

Honesty

People who are honest reduce the need for expensive security systems. Living, working, in short, being in a secure place contributes to a feeling of well-being, ease. People who live in a secure place feel free to walk, jog, move around. They are more peaceful, less anxious. Living, working, and being in an insecure place is the opposite of all that.

Lower class neighborhoods tend to be insecure.

The incarnation rate among the lower classes has increased. Crime has become endemic.

Murray is less sure about integrity - doing the right thing, whether it is actually criminal. Throughout the history of this country American businessmen have been known to be cutthroats. But usually, Murray says, they do not cross the line into actual fraud.

Religiosity

Being involved in a church prepares people to be involved in civic activities. One supposes this is true even for people who grow up and leave the church. The ground work, laid at any early edge, continues into maturity. Statistics show that church goers are healthier (physically and mentally) than non-church goers.

Since 1960 secularism is on increase - and it is the greatest (as measured by church attendance) in the lower classes. (A recent study shows that the percentage of people who are declared atheists has increases in recent years especially in among people in well-developed countries in Europe and Asia.

COLLAPSE AND HAPPINESS

The new lower class, because of failure to be virtuous as measured by the standards of the founding fathers, is in a state of collapse. That is what the numbers show - contrasted with the numbers of the upper class who seem to be doing well. Beyond a basic standard of living, it is not so much a matter of money, although that is a factor, but of more intangible things.

Happiness is one of those intangible things. Murray rates happiness as a product of...

- <u>Being Close to a Family</u> - Having intimate relationships with others, having a sense of belonging.
- <u>Having a Vocation</u> - Being satisfied in one's work - becoming self-actualized (being all you can be. According to Abraham Maslow's hierarchy of needs theory self actualization is the final level of psychological development that can be achieved when all basic and mental needs are fulfilled. Actualization is when is when the full personal potential takes place.).
- <u>Faith </u>- Believing in something larger than yourself. Not feeling alone in a very large universe.

HAPPINESS AND SELF RESPECT

Murray regards self-respect - which comes from being responsible for yourself and others - as the starting point for becoming happy.

Important in all the factors that contribute to happiness is employment and work – which derive from jobs. Murray does not mention that jobs are more available for members of the upper classes, especially those who do knowledge work. In an asymptotic curve plotting demand for all jobs against the availability of all jobs, the jobs and resulting happiness/unhappiness curve might apply primarily to the upper classes. However at some point the unhappiness of the invisible lower classes might overwhelm the happiness of the upper classes, resulting in a precipitous fall of the unhappiness curve.

Without self-respect it is hard to have intimate relations with your family or anyone else.

Self-respect, having a vocation and becoming self-actualized, work in a feedback relationship. Without self-respect you have trouble having faith in yourself and without some faith or regard for yourself it is difficult to have faith in anything larger than yourself.

And without self-respect and all that engenders, you cannot live according to the virtues of the founding fathers.

Lack of self-respect seems to be more of a problem among the lower classes (who seem to be regarded by Murray as a monolithic block - for instance he doesn't not consider that the mother who works two jobs and has respect for herself and her children). People in the lower class have come to rely on government for the basics - food, shelter, health care. The upper classes have enough money to avoid dependency on the government and the problems that ensue - at least not to the same degree.

EUROPEAN MODEL AND SELF-RESPECT

Murray says that the seeming rewards of the European model are illusory.

At this place in his book Murray acknowledges that he is now arguing his particular point-of-view, acknowledging that others might not agree with him. He discusses the European model social system arguing that this system will lead us to ruin. He admits that the European model seems attractive. Everybody is guaranteed a minimum standard of living at no effort. Medical care is available for everybody. It seems like a utopia where anyone would want to live.

However, according to Murray the rewards provided by the European model cannot go on forever. And even if they could it would still be a bad thing. In a welfare state personal responsibility is minimized. Not having a chance to do for themselves results in a loss of self-respect. And without self respect, as described in the preceding you cannot be happy.

You've got (in my words) European soul disease - angst, ennui and a too-keen sense of the absurd.

TWO ALTERNATE FUTURES

Murray sees two alternative futures - one in which the European model persists and the coming apart/collapse happens, and another in which there is a return - across all classes - to the virtues of the founding fathers.

Continuation of the European Model

Murray has already noted some basic problems with European model. The welfare state - by doing everything for people - leaving them with no responsibilities - takes away their self-respect. This in turn affects the person's ability to have close family relations, to have the will needed to pursue a vocation, to turn faith in one's self into to faith in something larger than oneself.

If the European model persists Murray also sees a decline of the Upper classes as well as the lower class. He already sees this in the behavior of some younger members of Upper classes. He complains that there is no sense of seemliness. People openly use language that was once the sole province of the lower classes. Some upper class youth wear tattoos (the horror).

Among adults (no doubt liberal adults) there is a hesitation to pass judgment on anyone for fear of being incorrect - or not having a clear sense of right and wrong and who is responsible for what.

He also notes that the obscene salaries of some privileged members of society are also unseemly.

Result: Coming Apart – Collapse

If the European model continues to hold sway and if the elite become selfishly self-absorbed and no longer perform their duty by setting a model for the rest of society the center will not hold and mere anarchy will settle – initially at least - on that part of the world. In other words self-absorbed European children, living the moment of immediate gratification will not be able (or willing) to run society and to handle the disasters that ultimately arise. This will be the coming apart – the collapse.

The Other Alternative – Learning from the Collapse of the European Model

Murray says we will see the failure of liberal beliefs when the European Model collapses. In asymptotic terms, the coming apart in Europe can be viewed as a local collapse. World-wide, the happiness curve might start to shake before the local collapse spreads. Perhaps it will be the local European collapse that provides the requisite shaking.

One key liberal belief is that all aspects of human nature can be changed for the better - that everything can be fixed. Murray (and he says, the founding fathers) understood that some aspects of human nature cannot be changed (the need for self-respect that stems for the satisfaction of taking charge of one's own life). Murray and the founding fathers say that government should stay out of these areas. '

Liberals believe that all people are equal in every respect - or at least can be made equal. The ditch digger has the potential to become the president of a bank and should be treated as such - anything else would be unfair. (Conservatives say that everyone should have equal opportunities and what they do with those opportunities - based on their abilities) is up to them.

Liberals believe that people are not responsible for what they do, They cannot be held accountable for their actions. The fault lies in bad genes and/or a bad environment.

Murray says that science will disprove the liberal beliefs that underlie the European model and in contrast will prove that...

- People enjoy themselves best when performing at their limits - which are different for each person. (In some physical fitness classes people are encouraged to go till they drop - with the understanding that the dropping point is different for each person.)
- People are not the same. Genetics, age, class, sexual persuasion, and much more are all different. These differences will produce different outcomes - in cognitive and physical abilities. Government cannot pursue a one size fits all approach. Fairness does not apply to these differences. There is nothing government can (or should do). In the movie Master and Commander which seems to accurately portray life on a British Man of War in 1805 there is no question about class distinction. Everybody seems to know and seemingly accept his place.
- Although people might not have free will (whatever that is) society functions best when people are treated as if they do have free will. (Murray believes that science will prove that we do have free will - at least in a "deep neurological sense".)

Murray says it will become obvious that the economics of the present bureaucratic system is ridiculous. People will understand that the same goals could be archived with less money.

Murray says that the first three Great Awakenings ushered in new eras. He thinks that the fourth era will renew and reaffirm our allegiance to the American Project.

Murray believes that we will chose allegiance to the American Project.

MY ASIDE – BACKGROUND TO THE ASYMPTOTIC VIEW - TWO KINDS OF INTELIGENCE

Following is an essay in which I speculate on the possibility of two kinds of intelligence. This could explain differences in jobs and employment between the lower classes and other classes which might, I propose, lead to the collapse of the happiness curve.

Doing this work I discovered something significant. Murray suggests (hints at, alludes to the possibility) that members of the upper classes might simply be smarter than members of the lower classes. I wonder if he is implying that the two classes have different kinds in intelligence. Could this help explain the ascendance of one class and the fall of the other and the ultimate collapse of both?

I think so.

Suppose.

Comparing the Two Kinds of Intelligence...

The intelligence of the New Upper class and the Middle Upper class is cerebral. The upper classes, consisting of knowledge workers (doctors, lawyers, programmers, writers, latte drinkers, cruise takers, etc.) shuffle information. They are educated not trained. They deal with ideas and data - with representations of the physical world not the physical world. One assumes that these people excel in intelligence tests.

The intelligence employed by the ever-expanding Lower class and the ever-diminishing Working class is more physical. Physical intelligence means being able to perform physical tasks. It is employed by people who deal with the "real" world - with sweat and blood. The term skill might be applied. Tools might be involved.

Just as there are different levels of cerebral intelligence there are different levels of physical intelligence.

(At one end are doctors who obviously excel at both cerebral and physical intelligence. Such people really lie outside the scope of this discussion.)

At the true upper level of physical intelligence are skilled craftspeople - people who might be regarded as artists. In the middle are tradespeople - carpenters, mechanics - operators of complex equipment. Many of these people are also skilled. At the other end are people performing the most menial physical jobs. Although the term skill might be used to describe the work these people do (he is a skilled flipper of burgers) it doesn't mean the same thing.

(Of course, skilled craftspeople and tradespeople employ cerebral intelligence in their work. One needs to know a great deal to be a carpenter, a plumber, a mechanic, a seamstress, etc. However, these trades and crafts require physical skill to express cerebral intelligence and that is what the upper classes see and judge.)

Values Placed on the Two Kinds Of intelligence...

The abstract intelligence of the upper classes is prized in today's technical digitized world. This sort of intelligence is required to make the data driven world work.

Less prized (since the Industrial Revolution anyway) is physical intelligence. Although still needed to make the physical world work it is seems to be regarded as a necessary evil. This intelligence is not much in evidence in the digital world where the real action takes place.

Lower end physical tasks are more easily described by algorithms and are more easily automated. (Algorithms are logical descriptions of processes - tasks and jobs. Simpler systems lend themselves to description by algorithms and are more easily automated. Complex systems less so.)

Based on my non-scientific observation the number of cerebral jobs is increasing and the number of physical jobs is decreasing which explains the point of this distillation and what will drive the happiness curve to collapse.

Murray's solution...

Murray's solution is to get rid of the welfare state (although he does not explain how). For the sake of argument one could concede his position.

However considering the points raised earlier it is possible to go a step further.

Suppose...

Upper class people became upper because of the value placed on their sort of intelligence. They have been assured of good paying jobs. They do not need government help and have largely avoided the effects of the welfare state. They can afford to adhere to the values of the founding fathers.

The physical intelligence of the people who moved from the Working and Middle classes into the Lower class is no longer valued. Their jobs have been eliminated - downsized, automated, outsourced and insourced. The values of the founding fathers might not make much sense to people on the edge.

People in the upper classes tell people in the lower classes to just suck it up. However, suppose the people in the Lower class don't have the sort of intelligence needed to just suck it up. Of course they could be educated into that sort of intelligence. But we know that is not going to happen.

The world is simply passing these people by.

The Welfare state, conceding all of Murray's points, still might be necessary to keep people from starving.

Suppose.

COLLAPSE

Penguin Books 2005

ISBN 978-0-670-03337-9

Jared Diamond

Collapse is about infinities that have been crossed.

Jared Diamond discusses societies, ancient and modern, that have collapsed. All the collapses occurred despite efforts to increase supply. He also discusses a few societies that controlled demand and did not collapse.

SUMMARY

Diamond notes various factors contributing to collapses, but tends to focus on environmental issues. He offers a "five-point framework of possible contributing factors". The italicized asides indicate how these factors relate to asymptotic systems.

1. Environmental damage caused by people *trying to increase supply.*
2. Climate change – today caused by people, in the past by natural factors *which tends to reduce supply.*
3. Hostile neighbors who prevail when a society becomes weak – maybe because of one of the other factors *which tends to reduce supply.*
4. Friendly neighbors who become unfriendly or weak and no longer support the society (maybe the formerly friendly society has been weakened by other factors) trying to *which tends to reduces supply.*
5. How a society responds to its problems. Are they smart, perceptive, honest – or the opposite? (*Do they futilely try to increase supply or wisely try to reduce consumption?*)

Diamond does not say that western-style culture will collapse. But he does point out how past collapses have been caused by ignorance and greed. Generally no one admits that a collapse is happening when it happens.

But it happens.

POLYNESIA – FALL OF EASTER ISLAND
The effects of isolation and one-upmanship.

Easter Island is the place with the spooky statues thought by Erich von Daniken to have been erected by aliens.

The island was first settled toward end of a wave of Polynesian expansion around 900AD. When Captain Cook arrived in 1774, it was sparsely populated by a group of subsistence farmers who did a little fishing in leaky, poorly built canoes. There were no trees and all land birds except for chickens owned by the inhabitants were gone.

Until modern times no one believed that such people could have erected the statues much less sailed across a thousand miles of open ocean to get here.

What modern archeologists have discovered is that at one time the island had trees, birds and variety of other flora and fauna. At its peak there were as many as 30,000 people living in nine or so communities that cooperated in some respects and competed in others.

The society ruined itself and the environment because of unrestrained consumption and finite resources. Competitive impulses, which could have been vented on neighbors if there had been any were turned inward. In particular the people competed (and consumed) in the creation of statuary – resorting to a one-upmanship game of my statue is bigger than your statue. The geography of the place is problematical to begin with which allowed the competition to destroy an already the fragile environment. At some point the island was no longer able to sustain the people and the population crashed –staying that way until the arrival of the Europeans.

Diamond notes that in some respects our modern globalized world is like an island – that the same pressures which brought down Easter island could bring us down Island Earth.

POLYNESIA – FALL OF OUTLIER ISLANDS
Ending up with nothing to eat and no place to go.

The outlier islands of Mangareva, Henderson, and Pitcairn in Southeast Polynesia were settled during the wave of Polynesian expansion around 800 A.D.

At first, societies prospered. Populations grew. Trade between islands compensated for the lack of resources on a particular island.

But after hundreds of years of growth, populations exceeded the capacity of the land. Islands could no longer produce enough to support themselves, nor anything extra for trade. Civil wars ensued. People starved, became cannibals. The land was devastated. A collapse happened.

Advantages and Disadvantages of Each Island

Each island offered advantages and disadvantages.

- Mangarvea, the largest, was capable of supporting the largest population. Except for a lack of fine grained stone suitable for making quality stone tools it had most everything the colonizing Polynesians needed.
- Pitcairn (the island on which the mutineers from the Bounty settled) was more problematical. It was small with steep hills that dropped directly into the Pacific. There were no surrounding reefs and little flat land for farming. Fishing in the deep surrounding waters was poor. However, from the Polynesians point-of-view Pitcairn did have saving graces. It had the raw materials – volcanic glass and fine grained basalt – for making high quality stone tools.
- Henderson, the most remote of the three islands was the most problematical. A coral reef thrust 100 feet above sea level by random earth movement, Henderson had no reliable sources of fresh water, no land for farming, no big trees for making ocean-going canoes. However, it did have abundant sea life in the shallow waters surrounded by the former reef. There were turtles and huge numbers of nesting seabirds. This was its appeal.

Trade Between Islands

The three islands constituted a regional trading economy (a miniature version of today's global economy). Mangarvea, the largest member of the trading group was in turn part of larger trading economy extending thousands of miles across the South Pacific into the Marquesa and Society islands. The islands were interconnected economically and socially. Trade, especially for the smaller islands became essential for survival.

Fall of Islands

Capitalizing on complementary advantages and disadvantages, islands across the South Pacific traded from around 1000 AD to 1450. Populations grew; economies prospered. However, these populations expanded into environments that had developed in the absence of humans. Animals, which had no defenses against humans, were decimated. These food sources were lost. Arable farming land was not able to make up the difference. Further, large trees suitable for building ocean canoes were cut down. Without canoes, inter-island trade was not possible – even if the trips had been worth the effort.

Consequently, in course of a few generations these societies collapsed. *Infinity was crossed*. People no longer lived on Pitcarin and Henderson islands (at least until the mutineers from the Bounty settled Pitcarin).

And that is what the Europeans found several hundred years later.

AMERICAS – FALL OF ANASAZI

Human environmental impact, climate change (natural), trade patterns, and a loss of faith caused infinity to be crossed and a collapse to happen. .

Anasazi civilization in the Chaco canyon region of southeast U.S. flourished between 600 and 1100 AD. During the peak of their culture, the Anasazi erected Great Houses of 600 or so rooms five or six stories high in the sides of canyons. These were the largest buildings in North America until the middle of the 19th century.

The Great Houses were the visible representation of a complex social system. Development and power were concentrated in Great Houses supported by farming and hunting taking place along the perimeter.

The Anasazi Great House culture was able to develop and survive because there were pinon trees for construction and a supply of water available most years. However, as the inner wealthy tier of society grew and prospered, the outer perimeter of poorer support villages expanded further and further into outlying areas. Nearby forests were decimated and water management practices destroyed farmland. Resources could not keep up with demand.

The system worked for a time. During good years when water was available populations grew and demand increased. After living many generations in such a manner, the people could not return to a simpler more sustainable style of living.

Consequently after the system was stretched past its limit the society could not withstand a draught (a Black Swan which caused infinity to crossed and a collapse to happen. Things then quickly fell apart. Groups fought over scarce resources. Cannibalism became common (not just as a ritual but as a way of obtaining food).

By 1200 essentially no one was left in Chaco canyon. The remaining Anasazi moved out to merge with other people.

Diamond said that four of the five collapse factors were involved in the fall of the Anasazi.

- Human environmental impact due to tree cutting and water management.
- Climate change - (natural not man made in this case).
- Internal trade patterns (which fell apart at the end).
- Religious practices which helped hold the society together for hundreds of years but which failed at the end.

Diamond said that a fifth factor, external warfare, was never an issue – although toward the end internal conflicts became violent.

AMERICAS – FALL OF MAYANS
It was population growth, deforestation, loss of arable land, droughts, and bad rulers.

Mayans occupied Mexico's Yucatan Peninsula and nearby areas of Central America. The region is about 1,000 miles north of the equator. There are irregular rainy and dry seasons. Farming is possible but unpredictable.

Mayan "classic" civilizations began around 250 AD and collapsed around 900AD. At its peak Mayan civilization was the most advanced of the three principal Mesoamerican cultures – which included the Aztecs of the Mexican Valley and the Incas of the Andean region of South America. The Mayans developed writing and a calendar.

At one point Mayan population numbered in the millions. By the time the Spaniard Cortez arrived in 1524 there were about 30,000 Mayans.

Mayan civilization was characterized by kingdoms containing cities with populations numbering in the tens of thousands. Nobles and other elites occupied the top strata of Mayan societies. The implicit contract between the classes was that the upper class – who talked to the gods and managed rituals – would ensure regular rains and good crops. Should that contract be broached, rebellion might ensue.

Warfare occurred between all levels of Mayan civilization - between kingdoms, between cities and between cities and the kingdoms of which they were a part. Wars increased in severity and frequency as the great collapse approached.

Geographically warring parties were generally nearby – within several days march. Diamond says this was because corn, the primary food, did not have much nutritional value – especially protein. It was not possible to carry enough corn to sustain an army far from home. Diamond contrasts corn with the more nutritious grains and other foods available to Europeans and to Aztecs and Incas. Having to rely on corn and not having large animals for carrying materials prevented a single strong ruler from conquering the entire region and eliminating the smaller wars.

Diamond includes the story of the Mayans because he wants to show that a large relatively developed culture can collapse – not just the smaller cultures examined thus far in his book.

He poses five reasons for the Mayan collapse:

1. The population exceeded the capacity of the supporting environment – in Malthusian fashion.
2. Deforestation and resulting hillside erosion reduced the amount of land available for farming.
3. Ever increasing warfare created regions of no-man's land between fighting parties, which further reduced the amount of land available for farming.
4. The most severe draughts that the Mayans had ever seen happened.
5. Short-sighted rulers, intent on wars and personal gain ignored the problems.

Note: These problems are an example of the main theme of Crossing Infinity – that the mismatch between stuff and need and/or greed) will result in collapse.

THE NORSE - ORKNEYS, SHETLAND, FAEROES, ICELAND, VINLAND

In some places being a Viking worked; in some places it did not.

Diamond recounts collapses, near collapses, and successes by Scandinavians/Vikings from around 800 AD to the present day.

Using shallow-draft, sail-and-row ships capable of crossing oceans and going up rivers, the Vikings started out as traders. Rich residents in Europe and Mediterranean prized the furs and beeswax from Scandinavia. However, according to Diamond, the younger brothers of older successful traders initiated the practice of taking rather trading. This is when the Scandinavians became Vikings, the traders became raiders.

In the beginning Viking raids were highly profitable because many of their victims where rich materially but weak militarily. The lure of easy wealth pulled younger Vikings economically. Population pressures pushed them geographically (perhaps it was the scarcity back home of women for marriage). However this period only lasted a few hundred years. Strong kings emerged in England and Europe. By 1066 when William the Conquer defeated the English at the Battle of Hastings, the Viking raids were largely over. (King Harold, who William defeated, was, according to Diamond, worn out from having recently defeated the Vikings in another part of England.)

Settlement of islands began during the raiding period. Sometimes the Vikings got lost and stumbled on uninhabited lands (Faeroes around 800 AD, Iceland around 870 AD). Other lands – the Orkneys, Shetland and Vinland (North America) were deliberately sought.

As they did in other conquered lands in Europe and England, Vikings married native women, adopted the customs of the conquered people and settled down, eventually blending in with the population.

The closer settlements were to Scandinavia the more likely they were to succeed. The survival of the Orkneys, Shetland and Faeroes (all islands North of Scotland) was never in serious doubt.

The survival of Iceland was problematical all along. Although knowledgeable farmers and herders, the Norse settlers misjudged the ability of land to support them. The soil appeared thick and rich but was in fact light and insubstantial. When exposed by destructive farming practices and marauding domestic animals (e.g., sheep), it was easily blown away by the near constant winds. Further, because Iceland was so far north, native vegetation including trees grew back slowly even where the topsoil had not been blown away. By the time the settlers – after multiple generations – figured out what was going on it was almost too late. The people survived by becoming very conservative in their handling of the environment. Some years individuals starved but the society as a whole endured to the present.

Vinland consisted of Newfoundland, Nova Scotia and the New Brunswick region of Canada. Viking colonies were established around 1000 AD. They lasted for about 10 years. It wasn't the inability of the land to support the settlers but the presence of hostile natives – who might have been made hostile by the tendency of the Vikings to kill them.

The next section of Collapse is devoted to Viking attempts to settle Greenland, a saga more relevant to our times.

THE NORSE IN GREENLAND - RISE
Getting by in spite of themselves.

Although Norse settlements in Greenland never exactly "flowered", they did survive for about 500 years, from 984 AD until the 1400's.

But survival was never easy.

Farming was only possible on land at the head of two fjords, removed from the salt spray of the coast and the ice and rock of the interior. Growing seasons were short. The climate was cold.

A big problem was the Norse themselves. Like all the other people described in Diamond's book, they cut down most of the trees and over farmed/overgrazed much of the land. Like many other societies they persisted in violent warlike habits, killing each other or the natives at the slightest provocation – for fun and/or profit.

The Norse also remained Eurocentric, refusing to eat much of the seafood that thrived in nearby waters, dressing in styles not suited to the cold climate, disdaining to learn from the Inuit (Eskimos) with whom they shared (and fought over) the land. They expended large amounts of scarce wealth trading for European luxury goods - spending that could have been better used for survival.

As described next, their attitudes plus a change in the climate did them in.

THE NORSE IN GREENLAND - FALL
Failing because of themselves

Norse settlement in Greenland ended around 1400 AD.

Diamond compares the lifestyles of the Norse with those of the Inuit. Although the Inuit where much more sophisticated at living in that region they were regarded by the Norse as "skraelings" or wretches. Norse euro centricity and Christian prejudice against pagans prevented them from adopting survival techniques presented right in front of their noses.

Diamond goes to some pains to note that we are looking at the issue with the benefit of hindsight. Given what they knew at the time, many decisions by the Norse made sense (using turf for walls and roofs for example). In any event, Diamond contends that many Norse decisions were no more suicidal than decisions we are making today.

Four factors sealed the fate of Norse settlers in Greenland...

- Climate fluctuation – the Norse settled in a warmer (relatively speaking) period and did not learn early on how to handle extreme cold. When the settlements finally collapsed in 1400, Europe had entered a period called the Little Ice Age. (In Nassim Taleb's terms, this would be a negative Black Swan which in my terms would drive the asymptotic curve upward.)
- Cultural predispositions – the Norse came to Greenland thinking of themselves as Christian Europeans first and Greenlanders second. Early on this attitude gave them the strength to persevere and survive but in the end left them rigid and incapable of adopting new ways necessary to overcome new difficulties.

- <u>Inability to learn from and deal with natives</u> – because of the cultural predispositions noted above it was impossible for the Norse to understand that as far as living in this region was concerned, the Inuit were smarter. Or if they did realize, they may have simply preferred to perish rather than adopt the ways of pagan wretches.
- <u>Norse power structure</u> – the top-heavy hierarchical Norse social system was deeply conservative and did not encourage innovation. Plus there was a lot of killing going on.

OPPOSITE PATHS TO SUCCESS (BOTTOM UP AND TOP DOWN)

Making it work. In terms of the Stuff/Desire/Collapse theme of Crossing Infinity, these are all examples of controlling desire – which in some cases involves limiting need and greed. Need was generally reduced by limiting population growth – either by having fewer children or using various means to get kill off adults.

Diamond looks at three successful societies, contrasting them with unsuccessful societies.   Two of these societies employed "bottom up" management techniques, with control coming from local levels. One employed "top down" management, with control emanating from central authority.

BOTTOM UP MANAGEMENT

It worked in two little places.

This technique is typically applied by smaller societies – or by smaller units in larger societies (for example sometimes in the US by county and local governments, each with its designated sphere of control). It can work where everybody knows everybody, where people are generally homogeneous, share the same values and culture – where, as a result, they tend to get along (or if they do fight do not engage in wars of extermination; like Vikings and Southerners - just fun). The small groups also need to control the land on which they live.

Diamond describes the bottom up management styles of the New Guinea highlands and the Tikopia Island.

New Guinea

The New Guinea Highlands occupy the upland regions of New Guinea, a large island north of Australia. The highest peaks are 16,500 feet and covered in glaciers. Until the 1930's the area was thought to be an uninhabited rain forest. When the first planes flew over, the region's valleys were discovered to be densely populated.

Humans had occupied the highland region at such population densities for 1200 years.

Diamond does not exactly explain how the New Guinea highland people managed to sustain themselves for such a long period of time while other cultures were failing. Part of it might be that they were lucky. The soil is good and occasionally replenished by volcanic ash and dust blown in from China. The rainfall is plentiful.

The native society should also be given credit. Diamond says they are among the most curious and experimental people he has ever met. Although their bottom up decision making process seems clumsy and slow they will try anything – such as casuarina tree silviculture and other beneficial agricultural practices.

Until coming under the influence of missionaries and outside government agencies the people managed to keep their population growth in check . Some of the techniques included war, infanticide, contraception using drugs obtained from forest products, abortion, sexual abstinence, and extended nursing of babies. One supposes that wars, led by tribal "big men" also provided some degree of diversion as well as population control.

Tikopia

Tikopia is a small tropical island isolated in the Southwest Pacific. The land is so small that the 1,200 people who live here are never away from sound of the surrounding ocean.   The island has been occupied for 3,000 years.

Generally the same sorts of factors that favored the success of the New Guinea Highland people favored the people of Tikopia. The land is relatively fertile, the people are sophisticated and innovative in how they manage resources and their own population. They are also homogenous.

Crops are grown in a multi-story forest of cultivated plants, with different foods available at each level. In effect, what the people lack in acreage they make up in elevation. In 1600 every pig on the island was slaughtered, thus eliminating a threat to the land. Another indication of flexibility is the fact that in lean times the people will eat foods they don't necessarily like. (Unlike the Norse inhabitants of Greenland who would not eat seafood that might have saved them from starvation.)

Population was (is?) explicitly managed. People felt that it was wrong to produce more children than the island could support. The same population control techniques used by the New Guinea highlanders were employed with perhaps an added emphasis on suicides – both explicit and virtual (e.g., young and maybe old people took long ocean voyages from which they could not expect to return).

TOP DOWN MANAGEMENT (JAPAN)
Working in a big place.

Diamond begins his story of Japan with the end of civil wars that went on from 1467 until about 1603. The latter date marks the beginning of the Tokugawa era, named for Tokugawa Ieyasu the warrior who won the last civil war and centralized control of the country. This era lasted until 1853 when Admiral Perry in his "black ships" opened heretofore closed Japanese society to the West.

In the era of peace that followed the end of the civil wars, Japanese population exploded, expanding land use at corresponding rates. In particular, the forests on which the Japanese depended for construction and fuel were beginning to be destroyed.

The tipping point (or in terms used by other writers, the Black Swan event which altered people to the Great Disruption taking place around them) was the Meireki fire of 1657 when Japan's capital burned. Realizing that the country's forests were in jeopardy, the rulers instituted strict land use policies. Pressures on farm land were reduced by using more seafood products. Confucian ideals of thrift and reduced consumption became official government policies.

Like the Tikopia people, the Japanese also controlled their population, using the same techniques – prolonged nursing, abortion and infanticide. Diamond seems to indicate that these measures were as much individual choices as government policies. People voluntarily adjusted family size in response to the availability of food and other resources.

Although the isolation of the Tokugawa era ended with the arrival Admiral Perry in 1853, many of the ideals of the era survived, some to the present day.

MALTHUS IN RWANDA

Explained by an academic from times gone by – and by the Stuff/Desire/Collapse theme of Crossing Infinity. .

Thomas Malthus was a famous English economist and demographer. He published a book in 1798 in which he argued that because human populations expand exponentially (200. 400. 800, 1600, etc.) and food production expands arithmetically (the next increase is independent of the last) humans will always starve. (Weathers' note – most of collapses described in this section have been "Malthusian".)

Today Malthus is not so much in vogue. Many advanced countries have managed to keep their population growth in line with food growth (or vice versa).

However, Diamond notes that both supporters and detractors of Malthus would agree that we could have a Malthusian collapse if non-sustainable resources are used up.

A modern case in point is the African country of Rwanda

In 1994 the Hutu majority went a rampage that eventually killed an estimated 600,000 members (about ¾) of the Tutsi minority. This massacre was started by the Hutu military and then was willingly continued by Hutu civilians.

The massacre is usually described as ethnic genocide. Hutus who had been dominated by lighter skin, more European appearing Tutsis since colonial times, were killing to satisfy old grievances. Cynical leaders, anxious to eliminate rivals, took advantage of this ethnic hatred to incite mobs. In 1993 businessmen close to the Hutu leadership imported 581,000 machetes which they distributed to Hutus to kill Tutsis.

Diamond says all this is true as far as it goes. However, there were other underlying factors - environmental problems, land shortage and social disruption.   In the years leading up to the genocide, population expanded, forests were cut down, and land became eroded. There was also a drought.

Because of population increase, available land became scarce. In Malthusian fashion the people exceeded the capacity of the land to feed them. Societal norms collapsed. Because land was so scarce, many younger people had no prospects of marriage and family – in short no prospects of a life. (In terms of the theme of Crossing Infinity, the desire/need side of the equation was adjusted.)

Therefore, Diamond argues, a deeper underlying reason for the genocide was this: it was a means for getting rid of excess people so that limited resources could go further. The Tutsi, despised anyway, made a convenient target. However, even in regions where there were no Tutsi, Hutus killed Hutus. Hutu victims included rich people, troublemakers, members of rival militias, and people who were already malnourished or starving – this latter group constituting the greatest number of Hutu killed.

In summary, Diamond says this is what a Malthusian crisis looks like on the ground - people without land or cows killing people with land and cows - people whose children walk to school on bare feet killing people whose children have shoes.

SUCCESS AND FAILURE IN THE CARIBBEAN
A tale of one island and two histories.

The Dominican Republic and Haiti occupy the same island (Hispaniola, in the Caribbean, near Jamaica, Cuba, and Puerto Rico). From the air the two countries appear very differently. The Dominican Republic is green, like a proper Caribbean landscape. Haiti is brown and denuded, like a moonscape.

From ground level, Haiti is even worse. Its rivers are silted with soil washed down from depleted land. It is overpopulated with one third of the island's territory and two thirds of the people. Its governments are generally corrupt and ineffective, offering very little in basic services such as water, sewage, electricity, education, and health care. Haiti is the poorest country in the world outside of Africa. It has no involved middle class, just a small elite who live in isolated enclaves and dream of France

The Dominican Republic, although not a paradise compared to first world countries is a paradise compared to Haiti. It has had functioning (if somewhat corrupt) governments. There is an involved middle class. Forests have been protected (sometimes by soldiers with guns) in a large system of national parks.

Diamond asks, how has this happened? One country has collapsed or is on the brink. The other country is more or less OK.

Part of the reason is historical.

Although the island was discovered by Columbus sailing under the Spanish flag, Haiti was eventually dominated by relatively rich France while the Dominican Republic remained dominated by increasingly poor Spain. At first this meant that Haiti was richer. Consequently, wealthy plantation owners could afford more African slaves. However it also meant that when these slaves revolted (and killed most of their former masters) the population density of Haiti greatly exceeded the population density of the Dominican Republic.

The slave revolt also resulted in different land use patterns. After the former Haitian slaves became free they reverted to the style of farming they had known in Africa. Each family acquired a small plot of land for subsistence farming. Consequently, as the population grew, more and more trees were cut down to support this lifestyle. Loss of trees for farming and to make charcoal for cooking resulted in land erosion. Land erosion silted streams and reduced the capacity of the soil to feed the people.

The Dominican Republic experienced some land use issues but it retained a plantation based economy longer. That, coupled with a lower population density resulted in less damage to the forest and the land.

Another difference between Haiti and the Dominican Republic was their attitude toward whites and the middle class. Because of their history as slaves, Haitians were suspicious of white emigrants. For this reason, a functioning, educated middle class never developed. In comparison, the Dominican Republic welcomed emigrants, including educated whites, from all over the world. Consequently a somewhat functional middle class did develop in the Dominican Republic. This allowed an export economy to arise, which did not happen in Haiti.

A final difference between the two countries was their luck with rulers. The Dominican Republic was led by Raphael Trujillo from 1930 until he was assassinated in 1961. Trujillo fit the definition of an evil dictator. However, for various reasons he protected his country's forest and fostered the development of a relatively prosperous middle class (which he could milk – and whose people he could murder if they got out of line). Trujillo's successor was the nominally democratic Joaquin Balaguer. He was president off and on from 1961 to 1996. More democratic than Trujillo, he expanded public works, eventually winning praise from US presidents Carter and Reagan. Today 32% of Dominican land is in public parks. The per capita income is five times that of Haiti.

Haiti was ruled by Papa Doc Duvalier from 1957 until his death in 1971. He also fit the mold of evil dictator. His son, Baby Doc ruled from 1971 until 1986. Perhaps marginally less evil than his father, he was ineffective and corrupt. During the Duvalier period hundreds of thousands of people were killed by the government; most remaining members of an educated middle class fled the country, and little attention was paid to the infrastructure and the economy.

What does all this mean – what does Diamond make of it?

He says the story of Hispaniola illustrates how countries influence their own fate – that it is not just naïve ecological determinism. He might be saying that the two peoples got the countries and the evil dictators they deserved. Or perhaps he is saying that fate of the two countries was determined by their history – that the people individually and collectively had no choice but to be the way they were. Maybe Papa Doc worked with (or exploited) the social material at hand – and that material was the product of a slave culture.

(It is characteristic of writers like Diamond to avoid explicitly deterministic explanations for human affairs. They always leave themselves – and humanity – an out – some degree of freedom. (Nassim would say that these are random acts caused by Black Swans.)

LURCHING ELEPHANT – CHINA
This elephant can crush us all.

China is the elephant in the corner of everybody's room. Whatever way it lurches the rest of the world is likely to be bruised.

Because of China's unique form of authoritarian government, decisions are made quickly and the lurches come fast. Some of the lurches work out – population control, ban on logging natural forests. Some don't – Mao's Great Leap Forward and the Cultural Revolution.

In *The Great Disruption*, Paul Gilding believes that China, because of the miserable state of its environment will be the first country to experience the great disruption (when things start to come crashing down and a Mad Max world seems just around the corner). Because of its ability to move fast, Gilding also believes that China will be the first country to go on a war-time footing to fight pollution.

Despite all the dismal statistics (summarized below) associated with China, perhaps the most significant feature is its growth and increasing wealth. The Chinese people aspire to First World status. They want to drive cars, live in larger, well heated and cooled houses, and eat lots of quality food. If even a significant portion of the Chinese people achieve these ambitions (not even considering aspiring Indians, Brazilians, and others), it will be like adding millions – even billions of people to the world's population. The planet simply cannot handle this growth. Something will give first. Gilding's Great Disruption will happen.

Following are some of the most interesting China statistics that Diamond offers:

- China is the world's <u>largest producer and consumer</u> of coal, the largest producer and consumer of fertilizer, the largest producer of steel. (That list goes on.)
- China has some of the <u>most polluted cities</u> in the world, has some of the most eroded, damaged soil in the world.
- Its <u>fresh water</u> supplies are below world averages.
- Although logging of natural forests has been prohibited, China is <u>forest-poor</u> by world standards.

- Using <u>outmoded polluting technologies</u> to manufacture its products, China is an exporter of goods and an importer of pollution. Actually, some countries send their garbage to China to be dumped.
- As the <u>third largest consumer of wood</u> in the world (for products and fuel) China must import lumber from other countries – thus contributing to de-forestation in the rest of the world.

WHY?
Youthful minds want to know.

Diamond's college students always ask how could societies make the disastrous decisions that allow a collapse to happen?

In other words…

1. Why don't societies <u>anticipate problems</u> (are the people just stupid)?
2. Why don't societies <u>recognize problems</u> even after the situation becomes obvious?
3. Why, even after the situation <u>becomes obvious</u>, do some societies, (or the people who could do something) not even attempt a solution?
4. And why do <u>some solutions fail</u> (again, are the people stupid)?

Aside: The people generally aren't stupid – maybe venal, maybe short sighted, maybe besotted by religion or tradition – but not stupid .

Problem Not Anticipated
Sometimes people just don't see a problem coming. (Black Swans happen.) There is a failure of imagination. British colonists in Australia did not imagine what would happen when they introduced foxes and rabbits. The foxes killed off smaller native animals and birds (none of whom had genetically coded experience with these foreign predators). Rabbits ate everything and even though the foxes ate the rabbits they couldn't keep up. Another example is kudzu –which was introduced in the southern U.S. to control erosion and ended up covering a lot of landscape.

Sometimes societies don't remember past problems. Non-literate societies can lose information from generation to generation. Literate societies can simply forget what has been written. For example just after the 1973 oil crisis Americans briefly switched to more economical cars. After a few years they returned to gas guzzlers.

Sometimes people draw false analogies to past problems (According to Nassim Taleb, Black Swans render theories based on past events suspect). Remembering apparent lessons of WWI, the French developed the Maginot line – a series of fortifications positioned to repel likely German infantry attacks. Unfortunately the Germans bypassed the Maginot line with masses of tanks – which had only been used individually in previous war. The French generals were stuck in the past and failed to anticipate the future. They learned the wrong lessons.

Problem Not Recognized

Sometimes people don't see a problem even after it arrives.

There are at least three reasons for such failures of perception...

Imperceptible

The problem is literally invisible to the naked eye. For example, a region's soil nutrients – invisible to the eye – might be missing. Native plants growing in such soil can appear lush, disguising the fact that all the nutrients are locked in the plants. When settlers, not knowing any better, cut the plants down the nutrients go away. Crops planted by the settlers won't grow.

Or, consider greenhouses gases. They are generally invisible. So are pathogens.

Distant Managers

Decision-makers might be somewhere else. They don't know when things go bad. Diamond notes that Tikopian and the New Guinea highland societies were successful – at least in part - because everyone involved was there. The bosses knew what was going on.

Slow Trends

Slow trends can be hidden in up and down fluctuations. Noise obscures signals.   Consider global warming. One year the temperature is up; another year it is down. Much data had to be collected before scientists agreed that global warming is real.

Creeping normalcy (landscape amnesia) can hide trends. When things happen slowly enough we might not see the changes. Over the years an untended field becomes overgrown and nobody notices until one day somebody says didn't that used to be a field? Or a slum happens. Or a town becomes gentrified. Or a stream gradually becomes filled with silt from runoff. Or gradually all the trees in a forest are cut down and only the old people remember back when.

Solution Not Attempted

Sometimes solutions might not even be attempted - for various rational and irrational reasons.

Rational Reasons for Not Attempting Solutions

People or groups often don't do anything because rationally (ignoring any moral issues) it is not in their interest to do so. Diamond cites the polluting company who leaves an area before the damage affects them. They take the money and run.

This type of issue is sometimes referred to as the "tragedy of the commons", "the prisoner's dilemma", and "the logic of collective action". Generally it means (again ignoring any moral issues) that when people can – with little penalty - gain an advantage by behaving badly it makes no sense for individuals to behave well.

For example...

If everyone is waiting in a line (vehicular, pedestrian) and you have an opportunity to cut in then why not do it before other people get the same idea? (You might be especially inclined to cut in if you don't feel anything for the other people – or, better yet, if you resent them because of social difference – in that case cutting in provides positive satisfaction.)

Or if everybody shoplifts, maybe you should as well – before the store packs up and moves out of your neighborhood.

Or if your local economy depends on an ecologically damaging industry (logging, mining, etc.) why should your town pay the price for a national concern?

There are two obvious answers to such problems.

One is to force people and groups to do the right thing by regulating their behavior. Such regulations are typically backed by people with guns.

The other answer is for people to police themselves. Diamond says that this can only happen in homogeneous populations where people share common community values. One class cannot feel especially estranged from the class above it. And it helps if people care about something larger than themselves.

Note 1: This is the top-down versus bottom-up approach to government that divides liberals and conservatives. Top-down advocates accept that many people will not do the right thing, either because they have no free-will – they are basically animals, idiots, whatever – or because they are selfish and don't care. Bottom-up advocates argue that might be true, but everybody should be free to do the right thing and government should not get in the way. People should look after themselves.

Note 2: Charles Murray, the libertarian whose book Coming Apart is distilled earlier in Crossing Infinity, says the solution to such problems is to return to the virtues of our Founding Father. And the only way to do that, he says, is to get rid of the Welfare State.

Irrational Reasons for Not Attempting Solutions

In previous examples, no solutions were attempted because inaction made rational sense at least to some people. Irrational inaction happens when there is no rational reason for anybody to maintain the status quo but they do it anyway.

Religious convictions sometimes prompt irrational behavior. Diamond offers as an example the Easter Islanders who cut down all their trees to get logs to transport statues – objects of religious veneration. He also mentions the Greenland Norse whose shared Christian values helped them survive for centuries then prevented them from adopting new lifestyles needed for continued survival.

Secular beliefs (often held in conjunction with religious beliefs) can also get in the way of survival. Sometimes these beliefs start as a rational behavior and over the years became irrational. Sometimes it is not clear when or if there has been such a shift. The self-reliant, go-it-alone pioneer spirit which helped found the U.S. might not make sense in a complex urban society where people must live and work in close proximity. Many people – especially the people who believe in bottom-up control and free will would disagree. Top-down people – many of them – would regretfully agree.

Irrational reluctance to solve problems can also result from a focus on short-term issues at the expense of long-term views. The current willingness in Washington to sacrifice environmental regulations for a possible short-term bump in employment might be an example.

A final example cited by Diamond of irrational reluctance to solve problems is psychological denial. People just don't want to face issues. It is too painful. He notes a study of the attitudes of people who live below dams. People who live farther away from the dam were more concerned about dam failure than people who lived closer. Diamond contends that those people – looking up at the dam every day – professed that they were in no danger in order to stay sane (by being irrational).

When the Solution Doesn't Work

Sometimes people try to do something but the solution fails - for various reasons.

The solution might too complex at the present time. People might not be smart enough – may never be smart enough. For example we may never eradicate kudzu. Or maybe science can never get to the very moment of the Big Bang (our Big Bang anyway) or find the God Particle or figure out Dark Matter.

The solution might be perceived as being too expensive. For example all kudzu could be pulled up by a huge labor force. But that would be very expensive. (Some environmental problems viewed today as being too expensive might later be viewed differently.)

Some solutions might be too little to late (or just too late or too little).

Some solutions might be just wrong, the result of ignorance. Forest management policies that allow undergrowth to build up result in more and hotter fires. (This solution was once the result of ignorance, now it is more likely economic – clearing out undergrowth is too expensive.) Dunking and burning purported witches is generally wrong and ineffective.

CONCLUSION

Maybe so.

Like other writers on this subject (but unlike Nassim Taleb – or me), Diamond does not want to say societal success or failure is random - or determined by environmental or other factors. He prefers to cite courageous and/or farsighted leaders and courageous and/or farsighted people – leaders and people who can learn from history and face the future.

BLACK SWAN

Black Swan   how black swans bring it all down   Taleb

Random House 2007

ISBN 978-1-4000-6351-2

Nassim Nicholas Taleb

Black Swans, positive or negative unanticipated events, can result in sudden changes in the asymptotic curves that describe supply and demand relationships between processes.

Negative Black Swans can push an asymptotic curve nearly vertical, causing infinity to be crossed. For example a curve representing consumption could suddenly be pushed toward the vertical if a Black Swan caused either a sudden increase in demand or a sudden loss of supply.

Positive Black Swans can cause the curve to level out For example a Black Swan that caused a sudden drop in demand or a sudden increase in supply could cause the curve to become less steep, perhaps even to slope downward.

Taleb (who never mentions asymptotic curves) is on a man on a mission. Like the Ancient Mariner and other crazy people, he seems compelled to tell his story. However, instead of being cursed by a dead albatross, he is plagued by Black Swans. They lurk unseen (and unseeable) in the corner of every room, waiting to peck us or preen us without warning. In Taleb's view, the world is dominated by Black Swans.

SUMMARY
The underlying theme of Black Swan is the notion of Black Swans. These are events which to a greater or lesser degree are unpredictable. Black Swans affect things in several ways.

Predictions of the future are only accurate if the likelihood of Black Swans is low. It is likely that the sun will come up tomorrow. It is possible to predict with some degree of certainty that you will not live to be 150 years old; but, assuming your health is OK, and you are not engaged in risky behavior, it is difficult to predict whether you will make it to 70. Various sorts of personal Black Swans might occur.

A less obvious effect is the view one has of the past. Often we forget that Black Swans were as likely then as now. We often attribute cause and effect to events that were the result of random Black Swans. We do this because it is in the nature of humans to construct patterns.

Black Swans raise doubts about the inductive problem solving method, which says that evidence from the past can be used to predict the future (in science and philosophy induction has been largely replaced by deduction).

In Taleb's view some people understand this division of the world and some don't. Those who don't often see order and predictability when there is really randomness dominated by Black Swans, This drives Taleb crazy.

MEDIOCRASTAN AND EXTREMISTAN

People live and work in two different lands. They generally don't know it. A speculator, entrepreneur and free-lace writer work in Extremistan, a prostitute, carpenter and dentist in Mediocristan.

Mediocristan

Taleb call the part of the world that can be more or less predicted Mediocristan.

This is a land dominated by the middle. Extremes are physically constrained. Uncertainly is of the tame kind, fitting neatly under a bell curve (explained below). In Mediocristan most people don't get crazy rich or really poor. You pretty much get out what you put in – until you hit a wall. Effort and reward correlate - depending on how many teeth you fill, cars you sell, lines of code you write, johns you service, etc.

The size of human heads is a phenomenon of Mediocristan. Heads are physically constrained from getting too big or too small. The distribution of head size can be neatly captured on a bell curve.

The "uncertainty" of a series of coin tosses or the operation of a roulette wheel is also phenomena of Mediocristan. (Both can be charted of a bell curve.)

Extremistan

Taleb calls the part of the world that cannot be predicted – that operates under the sway of Black Swans - Extremistan.

In Extremistan much is uncertain. For example, we don't know what new technologies will emerge in the next 25 years and how they will affect us. We don't know if events will validate the views of Gilding, Homer-Dixon and Murray - that the world is collapsing and coming apart. We don't know if a positive Black Swan (like the discovery of a bountiful new energy source) will invalidate these predict ions.

On a personal level we don't know if our appointment with the doctor will be late by 45 minutes because a previous patient came in with an emergency (a Black Swan). We do not know if going to work tomorrow morning we will get hit by a bus. We do not know if a Black Swan will cause us to walk around a corner and bump into our next true love. A writer doesn't know if tomorrow will bring a long anticipated acceptance letter from a publisher.

Extremistan – as the word suggests - is a land dominated by extremes. Physical constraints don't work. Black (or gray) swans rule. Using the terminology defined below, wealth is scale-free. You can get a lot for a little effort (or nothing for a lifetime of effort - you never know).

Weather takes place in Extremistan. A butterfly flaps its wings in Africa and a couple of days later a tornado happens in Kansas.

Most project development occurs in Extremistan.

The bell curve does not accurately represent Exremistan (having Bill Gates in the room throws off the income curve).

Scalable and Scale-Free In Mediocristan and Extremistan

The term "scalable" describes a system that can grow as demand increases. A system might start out scalable until it reaches a limit – the dentist can only fill so many teeth, the lady of the evening can only service so many johns.

A scale-free system is characterized by high volume and ever-escalating scalability. Unless something happens it can go on forever. (But something always happens.) In scale–free networks a limited number of nodes have a disproportionate share of links. These nodes are called hubs. The world wide web, electrical grids, food distribution networks, etc. are scale-free. As noted in Upside of Down, scale-free networks tend to be more robust. However if one critical node fails (if a Black Swan happens) the whole thing can collapse.

Mediocristan is more-or-less scalable; Extremistan is more-or-less scale-free.

PROBLEMS THAT DRIVE TALEB CRAZY
Much of The Black Swan deals with problems that drive Taleb crazy. They include problems..

- with collateral damage
- in thinking
- with the bell curve
- with fractals, gray swans and black swans
- with induction
- with data
- understanding the past
- predicting the past

WARNING! THOSE WHO TEND TOWARD DEPRESSION SHOULD NOT READ THE FOLLOWING.

PROBLEMS WITH COLLATERAL DAMAGE
Taleb is especially concerned about the plight of artists, scientists, entrepreneurs, speculators - anyone who labors in Extremistan. It is an unfair place where reward hardly ever matches effort. It is ruled by Back Swans.

Those of us in "concentrated" professions look for a Big Score which might, but probably will never come. We are sustained by tricks and hope. Although good luck can get a person, a city or a language to the top of the heap, good luck doesn't always hold. Bad luck happens (Black Swans poop) and the mighty fall.

We pretend that we don't care. We say that we don't need respect and recognition. We claim only to care about the process.

But it is a lie; we suffer.

Taleb tells the story of Yevgenia. She writes a book, self-publishes it on the web. The book sits there for five years, then takes off. She finds a legitimate publisher; gains a lot of publicity, makes a lot of money. The book is a black swan.

PROBLEMS IN THINKING

Taleb says, "We behave as if Black Swans do not exist: human nature is not programmed for Black Swans." We have problems thinking about Mediocristan and Extremistan. We tend to get them confused.

PROBLEMS WITH THE BELL CURVE

People think they live in a bell curve world. Taleb says for the most part they don't.

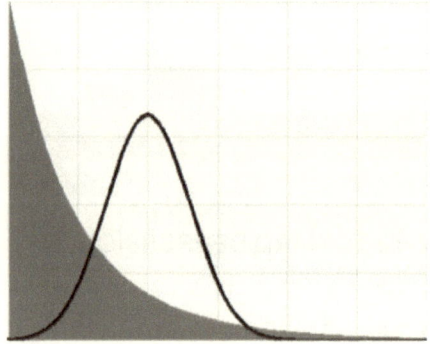

Figure 1. **A Normal Distribution (Black) Overlaying a Paretian Distribution (Grey).**

Bell Curve –vs- Pareto Curve

A bell curve (formulated by Johann Gauss in the 18th century - also called a normal, or Gaussian curve) shows the distribution of "random" events in a highly ordered reality. Most events take place in the middle of the curve which slopes on both sides to progressively unlikely events. The middle shows averages. Sigma is deviance from the normal caused by unlikely data that shifts the peak of the bell to the right or the left. The higher the sigma the greater the curve is shifted. A Pareto curve (formulated by Vilfredo Pareto in the 19th century) is a bell curve that has been shifted so much there is no longer an average. This curve is dominated by unlikely extremes – by Black Swans.

In Mediocristan, events can be chartered by bell curves. Things average out; stuff lumps in the middle. As things move away from the middle, the odds that anything different will happen increase exponentially. The more data that is included under the curve, the more perfect the curve becomes.

In Extremistan where Black Swans rule, events fall under a Pareto curve.

Legitimate Bell Curve Reality

Some aspects of reality can be legitimately described by a bell curve.

The bell curve applies where events happen independently of each other and have an equal opportunity of going one way or another - where questions can be answered by yes/no responses. Coin tosses, casino gambling, most games of "chance" are examples.

The bell curve also applies where events are constrained by physical reality from becoming extreme. For example the size of most living creatures is physically constrained between upper and lower limits (although a single spore organism can get really huge).

Confusing the Two Realities (A Taleb Rant)

Taleb rants about mistakes that result from the applying the bell curve to events in Extremistan......

- In the last 50 years, the 10 most extreme days in the stock market have resulted in half the returns. This means that five years of steady gains (or losses) could be offset by one crazy day (good or bad).
- The crash of 1987 would show up as way over on the edge of a bell curve.
- The modern portfolio theory (MPT - developed by two Nobel prize winning economists and the foundation for much theory regarding the market), is based on bell curve analysis and ignores power laws. .Taleb hates the MPT.
- Sometimes Black Swans work slowly. In the short term, the data might look like it might fall under a bell curve. However it only takes one wild (20 or 30 sigma) event to invalidate a theory based on bell curve analysis. The confirmation error is not an error until it suddenly is.

PROBLEMS WITH FRACTALS GRAY SWANS AND BLACK SWANS

Gray swans are the results of fractals (see below). Fractals are theoretically predictable but practically not. Many Black Swans are actually gray swans. We don't know when they will happen, just that they will. True Black Swans are completely unknown. They blindside us every time.

Benoit Mandelbrot

Benoit Mandelbrot formulated the concept of fractals. Mandelbrot was born in Warsaw in 1924 into a Jewish family that went to France in 1936 to escape the Nazis. In 1958 he moved to the US and joined IBM as a researcher. His work was founded on two ideas - that the world is loaded with self-similar phenomena and that much reality does not fit neatly under a Gaussin bell curve. The former belief led him in 1975 to coin the term fractal, which made him famous. The later belief led him to Nassim Taleb - or the other way around. (Seeing them together on TV is pretty weird, the placid old mathematician and the bouncing trader/philosopher.)

Fractals

Fractals refer to aspects of nature - physical stuff, events, processes - that are similar across all scales, big and small. Look at the basic shape of a mountain. It is pretty much the same viewed from a distance or close up. It's the same with branches of a fern - as well as crystals, snowflakes, coastlines, broccoli, lightning, etc. All are fractals. They are the result of fairly simple rules applied recursively - that is the same rule applied over and over to the same thing.

In a fractal system, the ratios of differences between large and small, tend to the same or close. Also there is no obvious limit to the size of the system. (There might be a limit, some point at which the system becomes predictable, or Gaussian; however, there is no way to determine what that limit might be.)

Fractals can be applied to visual arts, music, poetry, literature, any activity where there is a repeating theme or pattern.

Nonlinear Chaos

Fractals are examples of nonlinear, chaotic systems. These are systems that start out simple then over time, as rules are recursively applied, evolve into something much more complex. Such systems are sensitive to initial conditions. A small change at the outset results in major changes later. A simple fractal shape can quickly become a complex shape (that still contains the seed of the starting shape). A butterfly flapping its wings and creating tiny wind patterns in Africa can result in a hurricane in Florida.

Chaotic systems may appear to be stable for some number of iterations, then start to rapidly change. This is called the crossover point, when things start to get weird, when the system becomes "chaotic" (cross infinity?) The crossover point is related to exponent or power attributes of chaotic systems - which determines the difference between each iteration.

Taleb's point is that you can't get inside a chaotic system to see what these values and attributes are. Although such a system may in fact be deterministic, operating according to certain rules, there is no practical way to predict where it is going. Scientists who study complex systems have discovered certain patterns that might be universal across totally different external realities. Cool, Taleb might say, but you still can't predict what's going to happen. A system can start out Gaussian, become Mandelbrotian, then become Gaussian again.

He ends this section with a list comparing the two basic ways to approach randomness - the way of the skeptical empiricist and the way of the theory-bound Platonist. Generally, the former analyzes data from the bottom up (not the top down) and would rather be broadly right than narrowly right but broadly wrong.

PROBLEMS WITH INDUCTION

In Taleb's experience, theories based on experience – on induction - can always be undone by Black Swans.

Induction is reasoning from particulars to the general. You devise an explanation - a theory - based on experience.   Deduction is just the opposite - going from the general to the particular. You start with a theory and then come up with confirming observations. (Taleb does not have much to say about deduction.)

(Aside: The theories from which one deduces explanations of experience also have to start with experience. However, marshaling experience - in the form of empirical data - to support a testable theory seems more scientific.)

Parable of the Turkey.

There was a turkey who lived very well for 1000 days. Then on the 1001st day, Thanksgiving happened. The turkey got his head chopped off. If you had asked the turkey on day 1000 how things were going the turkey would have said, "Great".

These lessons emerge...

- Black Swans are relative to knowledge. Somebody smarter than the turkey would have been able to interpret the data – the experience – to figure out what was going to happen next.
- Once a particular black swan has happened we worry about the same black swan repeating itself. The remaining turkeys have been warned about Thanksgiving. But what about Christmas?
- Not all Black Swans happen quickly; they might take months or years to become apparent. For 1000 days the turkey lived in blissful ignorance.
- It only takes one contrary result (the turkey's 1001th day) to invalidate a series of positive results and invalidate any theories based on those results.

History of Induction Problem

Historically, the idea has been addressed by:

- Sextus Empiricus - A Roman doctor and philosopher living around 200 AD. The term "empirical" comes from his name. He said induction doesn't work because you need criteria to judge the truth of observations but that criteria itself needs criteria by which it can be judged which requires additional criteria etc. He also said that when judging the "universal" from the particulars you'd need to examine all the particulars to be absolutely sure of your judgment - which clearly would be impossible. In answer to all this, he advocated suspending judgment, neither affirming or denying truth. Be a skeptic. Let the facts speak for themselves - don't jump to conclusions based on incomplete evidence. Trust the empirical data. Don't generalize.

- Al-Ghazali (Algazel) - An Islamic philosopher (1058 - 1111) who reinvented skepticism for Muslims. He argued that what we see as cause and effect is actually God's will at work - and that it seems rational only because God is rational. He used reason and logic to propose a religious skepticism. He was refuted in the next century by Averroes, another Islamic philosopher who said there is no conflict between religion and philosophy. (He also said that existence precedes essence which is the foundation of existentialism.) Averroes influenced Christian and Jewish thinkers but did not supplant Algazel's influence on Muslims - which Taleb says explains current Islamic anti-scientific bias.

- David Hume - A Scottish philosopher and historian (1711 - 1776), one of the British Empiricists. Generally regarded (so Taleb says) as a foremost philosopher in the English language. He coined the phrase "The Problem of Induction". He has two problems with induction. The first is that just because some cause and effect sequence has always worked in the past (turkey's 1000 days) there is no guarantee that it will always work in the future. The second problem is pretty much the same as Sextus' first argument noted above. Inferring from examples of successful induction that induction works is using induction to prove induction. It is circular reasoning and bogus. Hume's answer is to trust our natural instincts to do intuition. In day-to-day life just don't worry about it. (Taleb says he takes just the opposite approach - in day-to-day life he applies skeptical risk-avoidance techniques and doesn't worry about the philosophical stuff.)

- Karl Popper - (1902 - 1994) Regarded as one of the preeminent science philosophers of the 20th century. Like Hume, he did not believe that just because something had always happened, it will continue to happen. You can make provisional theories. The turkey, if scientifically inclined, could have proposed a theory that said life for this turkey is good. Up until day 1000, the theory would have been true (provisionally). One day 1001, the theory would be proved false. Which was Popper's point. You can never use observed evidence (inductive reasoning) to absolutely prove that something is true - but you can absolutely prove something is false. This is an important point to Taleb, to which he returns and returns.

- Albert Einstien - (1879 - 1955) in his 1919 paper on "Induction and Deduction in Physics" said this about induction...

  "The really great progress of natural science arose in a way which is almost diametrically opposed to induction. Intuitive comprehension of the essentials about large complex facts leads the researcher to construct one of several hypothetical fundamental laws... both the fundamental law (axioms) and the consequences form what we call a theory... But he (the researcher) does not arrive at his system of thought in a methodical, inductive way, rather he, **snuggles to the facts** by intuitive choice among the imaginable axiomatic theories."

PROBLEMS WITH DATA
We look for patterns in data because it is very painful otherwise. We try to make reality conform to our ideas (originating in Mediocristan) when reality is more a product of Extremistan.

Following are some of our problems.

Domain Specificity
We don't judge evidence on its merits but on its context - how it is presented. Context confirms truth. Taleb calls this "domain specificity", which is responsible for these sorts of logical mistakes:

"Most terrorists are Muslims,

I saw a Muslim.

He is a terrorist."

We confuse "absence of evidence" with "evidence of absence".

Naive -vs- Negative Empiricism

Taleb uses the term "naive empiricism" to describe our propensity to look for evidence in what we already know. We search for confirmation. It makes us feel good. (The turkey says, "Well, I've had 1000 good days, why should I expect the rest to be any different?")

Taleb's answer to naive empiricism is negative empiricism as taught by Karl Popper. Although you can never be absolutely sure that your theory or proposition is true, just one negative result can prove it wrong. That's where truth comes from. Evidence is asymmetrical. One piece of negative evidence can offset a lot of positive evidence.

(Although you have seen hundreds or thousands of white swans, you can never be sure that all swans are white. However, if you have seen one black swan, you can be absolutely sure that not all swans are white. Or, even if your test says you don't have cancer, you really don't know. However, the one positive test that you do have cancer is proof.)

According to Popper, we should regard all theories as provisional. Don't look for what will prove you right but what will prove you wrong; it's a faster more certain process and you'll learn more.

Natural Instinct

Taleb says that deriving truth from evidence (using induction) is probably genetically built into us. Children seem to be born with the ability to inductively reason from particulars to the general and to know which truths to trust and which not to trust. (A child of two understands the truth of "dogness" and "catness".) In a more "natural" world (a more physical world, like Mediocristan) such reasoning is efficient and works. However, in a modern flat world that is built on abstractions (e..g, Extremistan) induction can't be trusted.
 There are too many feed-back loops, too many Black Swans.

(Hume proposed something similar when he said that although inductive truths can never be absolutely proven, in the real world the process seems built into us and should therefore be trusted - to which Taleb says, "nah – not in Extremistan".)

Tunneling

This is the tendency to focus on a small set of previous Black Swans, to become obsessed with known risks extrapolated from past experience. Inductively, we reason that the event will repeat itself and ignore the unknowable risks to come.

Narratives

The narrative fallacy is really a problem with information. Information needs (according to Taleb, "wants") to be reduced. Ordered, categorized information is compressed. It requires less storage room (in brains or wherever). Information entropy – the capacity for "surprise" or compression decreases. We had to invent categories in order to avoid being overwhelmed by details. We had to invent the forest so we wouldn't be blinded by the trees.

Taleb says that in the process of simplifying information, we lose sight of Black Swans. Since (before the fact) Black Swans cannot play any part in a story, we leave them out of the narrative. However, once a black swan happens, it can (and must) be fitted into our narratives. That is why we become obsessed with historical Black Swans - not as Black Swans but as events in a narrative.

Our Relationship to Narratives

We are biologically programmed to love stories. Finding patterns increases levels of the naturally occurring pleasure chemical, dopamine. Trying to look at unordered random data is physically painful.

We remember categorized information better than random data. Memory tricks involve creating easy-to-remember associations.

Narratives are therapeutic. They explain the chaos. (According to Mircea Eliade, explaining chaos is a prime motivation of religious man.)

Narratives interfere with our judgment of odds. Studies have shown that if people are asked to judge the likelihood of two events - one presented in an abstract manner and the other in the context of a story, we will most often pick the event described in a narrative.

Avoiding Narrative Fallacies

Although we can't avoid being drawn to narratives (that is part of what it means to be human), we can avoid narrative fallacies by...

- "Favoring experimentation over story telling."
- "Favoring experience over history."
- "Favoring clinical knowledge over theories."

And, as Taleb notes in other chapters, one needs to engage in non-stop bottom-up tinkering, to be open to luck, and to hang around where good luck happens.

PROBLEMS UNDERSTANDING THE PAST

Taleb says, "What we see is not necessarily all there is".

The past (yesterday, the day before, the last century, the last millennia) hides Black Swans from us and gives us a mistaken idea about what has happened; this is the distortion of silent evidence"

History

History is a formalized view of the past,

Taleb grew up in Lebanon. His childhood was shaped by the civil war occurring in this formerly peaceful place. The civil war was a black swan. It was history.

Taleb says this about history...

- History is a black box. We see what happens on the outside, but not the inner processes.
- History does not proceed smoothly, but in jumps as black swan events occur.
- Human problems with history result from a "triplet of opacity":
    1. We think we understand present history (when we don't - imagine historians sitting in front of a TV camera doing the "aw shucks" thing trying to explain the historical significance of some current event).
    2. Past history viewed from the present seems a lot clearer now than it was then (imagine the same historians explaining the inevitability of 9/11).

3. Learned people (the historians above) create overvalued facts and bogus platonic categories.

- Regarding the opinions of learned people (and the categories they devise), Humans tend to cluster around fashionable or accepted opinions. We look for patterns that reinforce the patterns we already perceive.

Categorizing is what we do and as always, it is at the expense of the real complexity in the world around us. Stuff is always more complicated than it seems.

Selective History

Winners write histories. Losers are often forgotten.

Looking backward at history to find evidence for our pet theories, we ignore other facts, other evidence. Over time, these facts, which do not get included in any compelling narrative that we or anyone else might write, get lost. They become "silent evidence."

When looking at members of a group, we concentrate on those who (1) have not been destroyed by negative Black Swans, and (2) have profited from positive Black Swans. For example, when looking for talented writers, we only find those who have been recognized, and recognition in such fields is largely a black swan phenomenon.

Taleb examines these questions (all of which relate to a selective reading of history)...

- Does crime pay? We might say, "no", based on reports and old-fashioned TV cop shows. But what about the successful crimes that never show up on reports? (In the series "The Wire" many of the bad guys - especially those near the top, lived to steal and kill another day.)

- Is beginner's luck real? It seems that way. However only the beginners who are winners continue to gamble (unless they become addicts which might require winning first). The losers quit and never get counted.

- <u>Does swimming cause a swimmer's body (or running a runner's physique)?</u> No. The picture gets distorted because those with such body types tend to do well in certain sports. Others don't do as well, drop out, and don't get included in the sample. The statistics get skewed.

- <u>Is God merciful?</u> For some, not for others. Suppose we are looking at Hurricane Katrina for evidence of God's mercy. We only look at those who have been miraculously saved, not those who have been miraculously destroyed.

Fallacy of the Lucky

Lucky people, being winners, tend to survive and write the stories. Narratives favor the lucky. (My aside - Stories are never told about the expendable extras moving in the background along the passageways of the Starship Enterprise. They don't exist.)

Taleb says that we have an "illusion of stability". We don't appreciate the luck that has allowed us to survive. We don't recall all those incidents when we nearly died, where our continued existence depended on toss of a coin - heads you live, tails you die.

Risk takers who tell brave stories about their survival have just been lucky. Those who didn't survive don't tell stories.

Fallacy of Optimism

Some people, having arrived alive, maybe even successful, at some point in their lives, tend to be optimistic. Paraphrasing, Taleb says this about optimism...

- People mistake luck for ability and free will. We not as free or as able as we think.
- People point to the seemingly upward path of evolution as an example of the positive effects of risk-taking. The risk takers who survive to reproduce are somehow superior and result in progressive evolutionary flow. Taleb (like Stephen Jay Gould) says evolution is a series of accidents - Black Swans.
- Optimistic scientists point to the Anthropic principle which says that the universe is uniquely favored to result in life. If the various laws which govern the universe had been just a little different life could not have emerged. Taleb says that

there is no bias. We are simply judging the universe from the standpoint of survivors. If the universe had not favored our survival, we would not be here to claim that the universe favored our survival.

Fallacy of Cause

We want to believe everything has a cause, a reason. There might not be a neat "because", only a "just because". Or perhaps there is no because at all, only random Black Swans.

PROBLEMS PREDICTING THE FUTURE

Citing Yogi Berra, Karl Popper and others, Taleb notes that we just aren't very good at predicting stuff - the future being what it is and us living in Extremistan.

Focusing on Predictable Predictability

Taleb says, "We tunnel, that is, we focus on a few well-defined sources of uncertainty, on too specific a list of Black Swans (at the expense of others that do not easily come to mind)"

"Nerds" according to Taleb, tend to focus on managed, controlled, safe uncertainty. The uncertainty of games is an example. This is the kind of uncertainty that fits under a bell curve. It is predictable unpredictability.

Real life doesn't work that way.

He cites the example of Tony, the street-smart Brooklyn trader, and Dr. John, who deals in statistics. Both are told that a flipped coin has landed heads 99 times. They are asked what are the odds of getting tails on the next toss. Dr. John says, "Naturally, 50/50." Tony says, "Nah, no more than one percent. The game has to be fixed. A legitimate coin would never give 99 heads in real life. The odds are too great." Dr. John is stuck inside his box, trying to model life on games and bell curves. Tony, a creature of Extremistan, sees life as it really is.

Taleb cites another example of the martial artist who has been trained in a rigid discipline with strict rules. He loses a contest to a street fighter who only plays to win – who is loose and takes advantage of luck. (This is the essence of the Israeli self-defense technique, krav maga.)

(I recall a similar example from a film. In the first Indiana Jones movie, when confronted by a sword-wielding bad guy, Indy shrugs, pulls out his trusty Webley pistol and shoots the guy. The Webley is a black swan, perfectly reasonable in retrospect, but completely unexpected by the smug sword fighter. You do not take a knife to a gunfight.)

"Epistemic Arrogance"

Taleb calls the problem "epistemic arrogance". We think we know a lot more about prediction than we really do. In cases that really matter we almost always get it wrong.

That seems to be the way we are built. It's especially true of our ability to predict Black Swans - which shouldn't be surprising, since by definition, they will be missed in any prediction. Our predictive abilities only work in Mediocristan.

Taleb says that our arrogance has implications.

We constantly overestimate what we can predict and underestimate the opposite.

There is no effective difference between guessing about what you don't know and predicting. They tend to be one in the same thing.

The more details we know about something, the more likely we are to make bad predictions. The details obscure the big picture.

Experts

People who work in Extremistan and claim to be experts at predicting are by definition frauds (although they might not know it).

Experts tend to focus - to tunnel in - on narrow problems, which lead to self-deception.

When expert predictions fail, it always seems outlandish, unusual. In his defense, the expert will claim that the event was a freak occurrence. That it could not have been expected. To which the savvy person replies, "Well, duh! Of course it could not have been expected. It was a black swan." The expert probably does not even realize that in

Extremistan, the freakish, the unexpected, is the rule. It is normal. The expert is attempting the impossible.

Taleb cites an experiment performed by Philip Tetlock ("Expert Political Judgment: How Good Is It? How Can We Know?"). The experiment shows that experts are no better at predicting than the rest of us. Further, the more important the expert, the worse the success rate of the predictions.

Experts never have any perception of their success rate. And complicated predictive software tools (according to Taleb) are no better than seat-of-the-pants prognostication.

Projects

Virtually no project of any significance gets finishes on budget, on time.

Consider, Taleb tells us, the example of the Sydney Opera House. It was supposed to be completed in 1963 for 7M (Aus) but was not actually finished until 1973 and then for 104M.

In Extremistan, projects are scalable (or scale-free) - that is, they can be stable then go wildly wrong.

We fail to account for or measure error rates when predicting projects. For example, you would not want to cross a river that is estimated to be "on average four feet deep" without knowing the error rate of the estimate.

Knowing the Unknown

The most meaningful developments are often accidental and unappreciated at the time. According to Taleb...

- Copernicus didn't get into trouble with the church until 75 years after he died.
- Early on, the church didn't pay much attention to Galileo.
- Darwin wasn't a big deal in his day (I think Taleb got this wrong - or I am not understanding him).
- People hardly ever predict the stuff that has the biggest impact (computers, lasers, the internet - probably cars and

airplanes). Predictions are usually just extrapolations of ideas currently in vogue.

- You can't really predict the future without knowing the future.
- Aside: Nearly all wars start by extrapolating the lessons of the past and being wrong. WWI did not anticipate rapid fire artillery, machine guns. WWI did not anticipate massed tank attacks.

What these men said about knowing the unknown and unknowable.

- Henri Poincaré (1854 - 1912) the French mathematician who laid the groundwork for modern chaos theory was one of the first to claim that there is a limit to what we can know. Although we might know the initial state of a thing, in nonlinear deterministic systems, small changes are quickly magnified. These are the gray swans of Benoit Mandelbrot described later in book. (In theory, even a chaotic, nonlinear system is predictable if you know enough - although practically, you can never know enough.)

- Karl Popper, as well as Poincaré and the economist Friedrich Hayek argued that events taking place in the social world (as opposed to the world of hard science) are basically unknowable. You can't apply the tools of hard science to such social phenomena; the data are too big, too fuzzy.

WHERE DO WE GO FROM HERE?
So how do we survive in Extremistan?

Don't Trust Planning

Because social phenomena is dominated by Black Swans and basically unknowable, Taleb does not trust central planners. He does not trust big governments and other big institutions. Companies (and their CEOs) are successful not because they are good planners but because they are lucky.

Don't Trust Top-Down Economics

If you believe that people have some degree of free-will and/or that people will sometimes act against their own best interest, then any activity involving people becomes unpredictable. This is opposed to the view of top-down economics which bases predictions on the premise that people do things to optimize their situations. Paul Samuelson is an example of the sort of economist who believes that the real world can be mathematically modeled. J.M. Keynes is an example of the kind of economist who believes that governments should interfere in economic matters.

Become an Epistemocrat

What Is an Epistemocrat?
An epistemocrat is someone who is very careful with facts, careful about what he knows. He doubts his own knowledge. That is his "way".

What Is Epistemocracy?
Epistemocracy is a (mythical) place where epistemocrats predominate. It is a place where everybody is aware of his or her own ignorance. In this place, people are confident about what is wrong. They learn from past mistakes. (The epistemocrat who once thought that buying a new car would turn his/her life around will never do that again.)

What epistemocrats Know
Epistemocrats know that history is often difficult, even impossible to unravel. For example, you could predict the behavior of a melting ice cube, but you'd have a hard time figuring out what had happened just by looking at puddle of water (unless maybe the freezer door was open and an ice tray had been removed, and even then you wouldn't really know).

Epistemocrats know that linear physical processes can be projected backward and forward in time but that nonlinear chaotic processes are something else altogether. Looking at a butterfly flapping its wings in Africa, you couldn't know that it would cause a hurricane in Florida (the word "cause" becomes inadequate). Just as you couldn't look at the hurricane in Florida and work backward to the butterfly in Africa.

Don't Take History too Seriously

Taleb (the ultimate epistemocrat) says that history can be a harmless pastime, with lots of neat narratives - we just shouldn't make too much of it. We shouldn't try to project history too much into the present or the future. We should be careful. (Does this mean that Taleb does not value the stories of collapse told by Jared Diamond?)

Don't Take Most Philosophers too Seriously

Taleb rants against philosophers - really anyone who ignores practical reality in the name of theories.

For example, he complains about those who use the Heisenberg Uncertainty Theory (which says a particle's attributes can be anything until you measure them) to support statements regarding ncertainty. He notes that at the macro level where we all live, particle uncertainty averages out. Quantum theory makes some of the most accurate predictions in science. It is very Gaussian. This, Taleb notes, is the essence of the Ludic Fallacy - confusing the controlled civilized randomness of the Gaussian with the wild uncertainty of the black swan.

We should, Taleb says, work outward from problems to theories, not the other way around. Philosophers typically go at it from the other direction.

Forget Fairness

Taleb Aside - Taleb says he "hates randomness, finds the word disgusting". Randomness makes the world unfair.

This is how unfairness works: Luck gives somebody an advantage (maybe the person takes better advantage of his/her luck). The advantage gets exaggerated because of the herd effect - the tendency of people to go with a winner. The exaggerated advantage gets even more exaggerated, and so on. Maybe the 1940's, 50's and 60's was a time of Mediocristan; maybe it was the Victorian era – maybe there was never such an era.

Unfairness is common in Extremistan. Skill and ability are supposed to be the deciding factors, but according to Taleb, are not. Arts, science, writing - all are governed by the power of the "cumulative advantage". It also applies to business - any activity where people benefit from past successes. In Extremistan, the rich get richer and the poor get poorer.

This is another term for how unfairness works. The more people do something the more they do something the more they do something. Some examples...

- <u>Words used</u> - The more we use words, the easier they are to use, the more we use them. Vocabularies get concentrated around smaller sets of words. .
- <u>City size</u> - Big cities get bigger. People congregate with other people.
- <u>Languages used</u> - English is the lingua franca of the world because as more people speak English more people are required to speak English.

Taleb notes the tendency in politics and economics to level the playing field. The little guys in the long tail try to assert themselves and restructure the pie (or the Pareto distribution curve). It is populism. However, Taleb says that even if the playing field were leveled economically, there would still be a pecking order, a hierarchy. The Pareto distribution would remain. Those at the top of the pecking order would still live longer, be happier. Life would remain unfair.

Fear Globalization

Interrelated world-wide networks are fragile. They can be taken down by a single black swan. According to the network theory of information entropy, network connections are concentrated in a limited number of nodes. Such networks tend to be resistant to random events, which are more likely to strike a poorly connected node (since there are more of them) than a well-connected node. However, if a black swan event does strike a well-connected node, it can destroy the whole network. Taleb cites the power blackout of 2003 as an example. He also worries about the concentration of big banks - resulting in curves with no long tails to take over if catastrophe strikes. (Homer-Dixon says the same thing.)

Find Hope

Using the example of Nero the Trader, Taleb says that is not bad to gamble on a big win in the face of continuous but small losses. However, this strategy requires that one has...

- "belief"
- "the capacity for delayed gratification"
- "the willingness to be spat upon without blinking"

One needs to think for the long haul.

Know the Difference between What Matters and What Doesn't

Finally...

- Know the difference between activities that result in negative and positive Black Swans. Minimize your exposure to the first and maximize your exposure to the second (e.g., put 85% of your money in T-Notes and the rest in a diversified portfolio of high-risk bonds).
- "Seize any opportunity, or anything that looks like an opportunity." Take advantage of good luck; it doesn't happen that often. Hang around where good luck happens. Go to parties. Socialize.
- Don't trust plans and predictions made by any organization. They don't know what they are talking about.
- Get in situations with asymmetrical rewards. Pursue risky ventures where the downside is tolerable and the rewards are great. Rather than worry about the probability of every act, perform a lot of acts.
- Don't tunnel. Don't obsess over the small stuff.
- Worry about what you can fix. Don't worry about the rest.
- Aggressively pursue opportunities when there is the possibility of great gain and limited pain. Otherwise, be careful.
- Choose your fate. Don't rush after trains you will miss.
- Don't avoid opportunities because of embarrassment.
- Marvel at the black swan of your own existence.

THE WORLD IS FLAT

Farrar, Straus and Giroux 2005

ISBN-13:978-0-374-29288-1

Thomas L. Friedman

The globally connected flat earth described by Friedman is a means by which asymptotic feedback loops can be happen. This affects this which affects that which affects that which affects this. Demand affects demand which affects demand etc. As the loop proceeds the asymptotic curve gets steeper.

Further, the flat connected earth is one way the connectivity described in Homer-Dixon's Upside of Down works. Connecting systems to one another allows things in one place to spread to another place. Good things and bad things can spread – inventions, ideas, wars, diseases, ecological disasters, and so on. The spread increases in a non-linear manner driving asymptotic curves to collapse.

SUMMARY

Friedman has written a book of clever stories and neat lists. He notes with enthusiasm how various technical and social forces have "flattened" the world, a process that has come to be known as globalization. Although acknowledging that some unfortunates have been squished in the great flattening, Friedman thinks it is generally a good thing. The other writers referenced in this book of books don't seem so sure. In a flattened world events in one place can quickly affect events in another place. A production failure at one end of a global supply chain can leave shelves bare at the other end of the supply chain. Animals and plants that co-evolved in one place can destroy an environment when transplanted somewhere else. Globalization makes such mingling more likely. Increased connectivity resulting from globalization also makes nonlinear feedback loops more likely. Black Swans can result. Asymptotic curves can collapse.

Friedman examines the potential and perils of the third great era of globalization - when people become globally connected across the world. The two previous eras (in 1492 when Columbus made his trip and in 1800 when traders roamed the globe) connected countries and companies.

Friedman says that the world has become flat as a result of globalization. Barriers between trade and communication have come down (been flattened). The playing field on which people, companies, and countries compete and play has been leveled. The world is no longer large and round; it is a small and flat. The people on the playing field are just as likely to be brown and Eastern as they are to be white and Western.

Ten forces, some social (such as the falling of the Berlin Wall) and some technical (such as the rise of internet and broadband communication) flattened the world. These forces converged with business and political developments to create what Friedman calls Globalization 3.0.

Of course, the flattening process is not complete. Many irregularities and rough spots remain. Nor do all people and countries profit equally. Some people, the sick, old, ignorant, and less agile are likely to get squished in the grand flattening. Nor is the process inevitable or irreversible. A grand old fashioned war could throw everything off.

Right now we are in a sorting-out period, figuring out what (and whom) gets flattened and what (and who) is allowed to remain somewhat irregular - perhaps for compassionate or sentimental reasons.

Friedman offers advice to individuals, companies, and countries on how to survive and prosper during the great flattening.

FLATTENING THE PLAYING FIELD

On a visit to the InfoSys company in Bangalore, India - seeing how this company does business across the globe, and how smart young Indians are doing work that Americans once did, Friedman had an epiphany. The world is flat, meaning that it is no longer large and round, but small and level. The field on which businesses and people play has been flattened.

The three eras of globalization are:

- <u>Globalization By Countries</u> - From 1492 (when Columbus made his trip) to 1800. This was the period when countries went global (although some countries and companies – like The East India Company and England – were indistinguishable in their aims). The era was driven by a desire for power and resources, and facilitated by technical and political developments.

- Globalization By Companies - From 1800 to 2000. This was the period when companies went global. The era was driven by multinational companies pursuing profit across the globe. The process was facilitated by advances in technology - especially in communication and transportation.

- Globalization By Individuals - From 2000 to the present. This is the period when individuals go global. Using the latest tools, people can compete directly across the globe. For example...

  1. A call center employee in India can manage a McDonald's drive thru in Kansas.
  2. A soldier in Kansas can manage a remote-controlled Predator drone airplane in Iraq.
  3. Contract programmers in India, Russia and the U.S. can compete across the internet for the same work.

  Globalization 1.0 and 2.0 depended on hardware; globalization 3.0 depends on software and hardware.

THREE FLATTENING FORCES
These are flattening forces which contribute to the current era of globalization:

1. 11/9/89 -sloshing capital and an operating system for the masses When the Berlin Wall came down it signaled an end to Soviet style planned economics and ushered in a grand new era of free-market economics. Liberated capital sloshed around the globe, funding a frenzy of trade. Everybody started to get connected - which was made easier when the number of Windows-equipped PCs reached a "critical mass".

2. 8/9/95 – internet for the masses Netscape browsers helped open the internet to the masses. The Netscape company, by going public, helped begin the dot com boom. The dot com boom, before it became a bubble and burst, prompted the

            installation of the broadband communication networks that
            made Globalization 3.0 possible.

3. <u>Internet backbone protocols</u> The development of underlying
   software protocols like XML, SOAP, and TCP/IP allowed
   systems to connect to systems across the internet. The
   playing field between applications was leveled. Company A's
   business rules could interoperate with Company B's rules.

In the mid 1990's a platform (a flat structure) began to emerge as a
result of the three forces listed above. The flatteners listed next use
this platform.

- <u>Open source communities</u> Open-source software like Apache
  and Linux empowered individuals and flattened the globe.
  Open sources of information and opinion, like Wikopedia and
  blogs, did the same thing.

- <u>India mines Y2K crisis</u> Using broadband networks laid down
  during the DOT Com boom era, India was able to take
  advantage of the Y2K crisis - when all the computer code in
  the world had to be examined line-by-line for date-related
  limitations. This set the groundwork for further, more
  sustained outsourcing work.

- <u>Offshoring meets outsourcing</u> China joined the World Trade
  Organization (WTO) in 2001, agreeing to trading rules
  imposed by the organization. This opened up a new era of
  offshoring - where entire manufacturing operations were
  moved overseas. (Outsourcing is the movement of jobs and
  tasks.) Both offshoring and outsourcing depend on the
  platform built in Events 1 through 3 listed above. Other
  developing countries followed suit.

- <u>Global supply chains</u> The creation of automated, just-in-time
  global supply chains made offshoring practical. (In an aside,
  Friedman notes that WalMart management prefers that

manufacturing be done locally, thereby giving local
employees more money to buy WalMart products. It is the
drive for lower costs and increased efficiency - by WalMart
and its suppliers, which determines where the manufacturing
goes.)

- <u>Equal access to information</u> Services provided by Google,
  Yahoo, MSN Web Search and similar companies gave
  everyone the same access to information. Everyone had an
  equal opportunity to become informed. The information
  playing field was leveled.

TRIPLES IN TRIPLICATE

The personal era of globalization was brought about by three factors:

1. The <u>coming together</u> of the flattening forces described
   previously.
2. The <u>development</u> of business practices (and business people)
   to take advantage of the flattening forces.
3. The <u>emergence</u> of billions of people who could play on this
   new playing field. (These people gained access to the field
   due to more liberalized economic and political systems -
   which resulted from the falling of the Iron Curtain).

These three factors were facilitated by three other factors:

1. The bursting of the <u>dot com bubble</u> (which, according to
   Friedman, actually accelerated the personal era of
   globalization).
2. <u>9/11</u> and the resulting War on Terrorism, which shifted focus
   from economic issues.
3. <u>Enron</u> and similar corporate scandals, which fueled the
   public's distaste for big business, and which made George
   Bush wary of appearing to get too close to Big Business.

SECULAR SECOND COMING

"Flattening the world" means removing the obstacles and frictions that impede the flow of trade, goods, ideas - and money. This is what globalization is all about.

Friedman notes that this is also what Marx and Engels wrote about in the Communist Manifesto in 1848. They said that capitalism and its flattening pressures would eventually destroy all nations, religions, and social systems - all barriers. Capitalism would end in a self-induced sorting-out process (a sort of secular second coming) when the historical process would finally unfold. There would be an economic Armageddon when the forces of labor would defeat the forces of capitalism.

Friedman's point is that we are now in such a sorting-out period. We are deciding (or the market is deciding) what frictions are desirable and which must go.

In this sorting-out period, we ask...

- In the interest of efficiency and profits, which jobs will stay and which jobs will remain?

- Will there be room for sentiment in the sorting-out process, or will all decisions be based on cold efficiency?
- Who or what will get squished?

IS FLATTENING GOOD FOR AMERICA?

Is free trade good for America in a flattening world?

Friedman says, "Yes." He believes that not all the inventions have been invented, that Americans can go on inventing (which is what we are really good at). He also seems to believe that increased consumption will go on forever, meaning that even if our slice of the pie gets smaller, the pie will get bigger. We'll still get more pie

Aside - This is not what Jared Diamond says in *Collapse*, what Thomas Homer-Dixon says in *The Upside of Down* and what Paul Gilding says in *The Great Disruption*.

If America competes well, jobs that are lost will be replaced by new jobs resulting from new inventions and ever-increasing consumption. (Or not, if you believe Diamond,Homer-Dixon, and Gilding.)

Aside - Richard Ricardo (1772 - 1823) was a British economist who said that free trade is good. Trading countries should specialize in what they do best. Everybody will profit.

SAFE WORK AND WORKERS
The work performed by some people can't be outsourced.

Unique People These are unique or famous athletes, businesspeople, actors, etc. Michael Jordon is unique. So are Bill Gates and Oprah. They can't be outsourced. (But not everybody can be unique.)

Fuzzy Work These are people whose work is so specialized that it can't be outsourced. High-level system architects and some technical writers are examples. They do work which can't be easily digitized. The rules for such work are fuzzy and ambiguous. (The rules for outsourced work must be describable in clear explicit language.)

Geographically Restricted Work People (plumbers, electricians, carpenters, cooks, prostitutes, nannies, etc) performing work restricted to a particular place are relatively safe. A nanny in Bangalore can't look after your kids in Charlotte. (However, a nanny from Mexico City who has come to Charlotte can do it.)

Adaptable People People who learn easily, who can adapt to new situations, who keep up with new technologies are usually safe. These people learn to do the work no matter how the work changes.

AMERICA'S ABILITY TO COMPETE
America's ability to compete and innovate in the new, flat world is affected by these factors:

1. Retiring Talent Many scientists and engineers are reaching retirement age. These are our innovators and creators - the ones who compete best in a flattened world. The number of such people coming into the system does not match those leaving.
2. Laid-Back Workers According to Freidman, American students are just not as ambitious as their foreign counterparts -

especially the brown ones (although we might rank up there with some of our more laid-back European cousins).

3. <u>Dumb Workers</u> Compared to our foreign competitors, American students are becoming less educated - especially in the sciences.

REQUIRED TO GET BY

The current situation requires the following:

1. <u>Honest Leaders</u> Political leaders need to explain to people what is going on. They can't hide behind easy populist rhetoric (decrying the evils of NAFTA, etc.) They need to convey a sense of urgency.

2. <u>Job Hopping Workers</u> In order to have lifetime employability (but not a single lifetime job - that era has gone) workers need help moving to the next job when the previous job ends. They need:

   - *Portable benefits* that can be carried from company to company (allowing the workers to follow the work).
   - *Opportunities for lifelong learning* People need constant learning and training to compete in a flattening world.

3. <u>Keeping the Peace I</u> Some social and economic safety nets need to remain in place - not just to be a "compassionate flatist" (Friedman's term) but to prevent unrest among displaced workers.

4. <u>Keeping the Peace II</u> Companies need to be socially involved - not just to be compassionate flatists, but to avoid social unrest.

5. <u>Looking After the Next Generation</u> People need to be better parents, to get their kids off the couch and away from TV's and Gameboys. Otherwise, the kids will not be able to compete in a flat world.

WHAT DEVELOPING COUNTRIES NEED TO THINK ABOUT

1. <u>Honesty</u> Countries (e.g., their leaders) need to be brutally honest about themselves - their problems, where they stand. They need to forget pride.

2. <u>Cheap Stuff</u> Toward the end of the first era of Globalization countries like China, Russia, Mexico, Brazil and India adopted a wholesale model of economic development. Using authoritarian powers, leaders pushed their countries into export-oriented, free-market economics. They made vast quantities of cheap stuff and sold it to the developed world, especially the US. This is the first level of economic reform, opening up trade and applying changes at the macroeconomic level.

3. <u>Efficient Stuff</u> There is always some less developed country ready to undercut competitors. In order to compete friction points need to be removed at each of these levels...
   - infrastructure
   - regulatory institutions
   - education
   - culture

   For example, regulations need to smart, not onerous or punitive. People need to be educated.

4. <u>Values</u> A country's culture (its values) largely determines its ability to get flat. If the culture does not value hard work, getting ahead, honesty, deferred rewards, it will not get ahead. Another factor is how well the country absorbs the ideas of other cultures. Insular, self-absorbed cultures don't fare so well in a flat world.

5. <u>All of it</u> How does one country get its act together and another doesn't? It seems to be a matter of culture, values, and political systems. Also it seems to be a matter of how education is viewed and how well educated the population is.

HOW COMPANIES COPE
Here are seven rules for corporate coping a flat world:

1. <u>Get smart</u> As the world gets flattened, more products and services get commoditized. Lots of people all over the world can build the same product or do the same thing (for

example, build flat screen TVs and operate call centers). Given equal quality, customers buy commodities based on price. Who's got the cheapest? The trick is to dig in. Find products to build and services to provide that can't be digitized - or turned into commodities. Depend on creativity and imagination - not just being able to provide the lowest price.

2. <u>Act big</u> One way that small companies can flower in the flat world is to act really big. And the key to being small and acting big is to take advantage of flat-age collaboration tools. Use the tools to build global just-in-time supply chains - to insource and outsource.

3. <u>Think small</u> Big companies can act small by finding ways for their customers to act big. Using the internet and automation, allow customers to do their own designs, create their own products. An example is LuLu, the web-based publisher of on-demand books. Anyone can be a publisher and sell books across the web (although as Nassim Taleb would note there is no guarantee that you will sell anything, since book publishing operates in Extremistan not Mediocristan).

4. <u>Collaborate</u>   Find your own core competencies and collaborate with others to outsource the rest.

5. <u>Know your strengths and weaknesses</u>   In Friedman's terms, a chest x-ray is an analysis of your company's parts. Figure out which of these parts make you the most money then outsource the rest.

6. <u>Outsource to win, not to shrink</u> The best companies outsource to innovate faster and more cheaply in order to grow larger, gain market share, and hire more and different specialists - not to save money by firing more people.

7. <u>Outsourcing is also for idealists</u> Some companies (often NGOs doing good works across the globe) and their "social entrepreneur" founders use flat-earth techniques to do social good.

PEOPLE LEFT BEHIND

For these people, the world remains unflat:

1. Too Sick Some people, especially in poor undeveloped countries, are just too sick to participate in the grand flattening. They are the people with AIDs and malaria, the people who are starving.
2. Too Disempowered Some people (in all countries - undeveloped, developing and developed) are too ignorant or powerless to benefit by the grand flattening. Some of these people might work in third-world sweat shops - the lowest places on an otherwise flat plain. They are regarded by the foes of globalization (old hippies, guilty rich people, guilty liberals, anti-imperialist in general and anti-Americans in particular) as the victims of globalization - the squished as opposed to the squishers.
3. Too Frustrated Some people, many in Muslim Arab countries, are too angry to join in the great flattening (although they might use its tools to spread terror). Some are religious fundamentalist who hate modernity. Some are angry youth who envy Western wealth. Humiliated by their lack of success, they hate the West as the perceived source of their humiliation.

DANCING IN A FLATTENED WORLD

Friedman tells the story of how his new Dell laptop came together in an elaborate supply-chain dance across the entire globe. This dance could only happen in a flattened world. And it could only be disrupted by a "good old fashioned war".

Aside – According Thomas Homer-Dixon and Paul Gilding a good old fashion ecological/economic collapse could also bring down the supply chain.

Friedman's point is that wars are less likely to happen when conflicting countries belong to a global supply chain. Being a part of a global supply chain is like having an oil well, except that this well taps into a pool of money. Getting into a war would disrupt the chain, which would cause the supply of money to dry up. He thinks this has been a factor in keeping Pakistan and India from having a full-scale war. (Friedman made a similar point in a previous book, *The Lexus and the Olive Tree,* when he noted that two countries that both have McDonalds will be less likely to fight a war.)

Aside- Mutated Supply Chains Al-Qaeda takes advantage of the flattened world to operate its own a global supply chain of terror. According to Friedman, a nuclear device in the hands of globalized terrorist organization would be "the mother of all unflatteners".

COMPARING 11/9 TO 9/11

11/9 and 9/11 offer two competing views of world flattening:

- 11/9 When the Berlin Wall opened up and when the number of Windows operating systems across the globe reached a critical mass. These were great flatteners.
- 9/11 When Osam bin Laden's global supply chain of terror brought down the World Trade Towers. This was also a great flattening event.

Both 11/9 and 9/11 were acts of imagination - one positive and one negative. Both were the fulfillment of dreams. Negative acts are caused by those who dream of the past. Positive acts are caused by those who dream of the future.

Examples of two groups who dream positively are:

- eBay and its open community of users Friedman calls eBay a "virtual republic comprised of people who have positive dreams". Everybody has the same opportunity. Your value is determined by your product and your customer rating.
- India and its 150 million Muslims India is the second largest Muslim country in the world. However, although they do not live in utopia, these Muslims generally don't join al Qaeda, generally don't become terrorists. They live in hope for their future in a flattening world.

- <u>Curse of oil</u> A major curse for people living in countries like Venezuela, Nigeria, Saudi Arabia, and Iran (and Russia?) is their oil. It allows totalitarian regimes to remain in power. By exploiting their natural resources they do not have to develop their human resources. For example, people change when they see examples of others like themselves changing. These examples encourage positive dreams of the future.

CONCLUSION

The world is flattening - for good or ill. Barring the disasters described in the preceding, it can't be stopped.